Carbohydrate Chemistry

Benjamin G. Davis

Lecturer at the University of Oxford and
Fellow of Pembroke College

Antony J. Fairbanks

Lecturer at the University of Oxford and
Fellow of Jesus College

OXFORD
UNIVERSITY PRESS

OXFORD
UNIVERSITY PRESS

Great Clarendon Street, Oxford OX2 6DP

Oxford University Press is a department of the University of Oxford.
It furthers the University's objective of excellence in research, scholarship,
and education by publishing worldwide in

Oxford New York

Auckland Cape Town Dar es Salaam Hong Kong Karachi
Kuala Lumpur Madrid Melbourne Mexico City Nairobi
New Delhi Shanghai Taipei Toronto

With offices in

Argentina Austria Brazil Chile Czech Republic France Greece
Guatemala Hungary Italy Japan Poland Portugal Singapore
South Korea Switzerland Thailand Turkey Ukraine Vietnam

Oxford is a registered trade mark of Oxford University Press
in the UK and in certain other countries

Published in the United States
by Oxford University Press Inc., New York

A catalogue record of this book is availaable from the British Library

Library of Congress Cataloging in Publication Data

Data applied for

ISBN 978-0-19-855833-0

6

Typeset by Newgen Imaging Systems (P) Ltd., Chemmai, India
Printed in Great Britain by
Antony Rowe Ltd, Chippenham, Wiltshire

Series Editor's Foreword

Carbohydrates (polysaccharides) are fundamental to Life. They are formed from combinations of sugar units (monosaccharides) and are involved in many essential biological processes: They act as an energy source, in biological signalling and recognition mechanisms and as basic structural building blocks controlling the architecture of Nature. For example, sugars are integral parts of DNA, RNA, starch, cellulose, chitin (insect and lobster shells) and cotton. Glucose, mostly in the form of its polymer cellulose, is the most abundant carbon fragment in the world. It is not surprising therefore that carbohydrate chemistry is a core topic in all undergraduate chemistry courses.

Oxford Chemistry Primers have been designed to provide concise introductions relevant to all students of chemistry, and contain only the essential material that would normally be covered in an 8–10 lecture course. In this Primer Ben Davis and Antony Fairbanks provide an excellent, easy to read and understand, account of carbohydrate chemistry showing the reader that this apparently complicated and daunting topic is simply the combination of elementary carbonyl chemistry with alcohol and diol chemistry. This primer will be of interest to apprentice and master chemist alike.

Stephen G. Davies
The Dyson Perrins Laboratory, University of Oxford

Preface

Carbohydrate chemistry is a central part of any undergraduate chemistry course. Often perceived as being a complex topic, it is, in fact, simple and consistent and can be mastered with a handful of key concepts. This book is divided into chapters that are intended to explain and illustrate those concepts. A good grasp of carbonyl and hydroxyl chemistry makes carbohydrate chemistry straightforward and clear. By drawing on existing knowledge and by relating the chemistry of carbohydrates to general themes of chemistry, we have attempted to dispel some of the myths of apparently strange reactions and unusual structure.

Much of the cutting edges of synthetic, medicinal and biological chemistry meet at the crossroads of carbohydrate science. While this book focuses primarily on chemistry, some of the vital roles that this key class of biomolecules play in Nature have also been explored from a molecular standpoint. Indeed the divide between chemistry and biology is now one that is quite rightly blurring and we hope that this book helps those from *all* disciplines wishing to explore this exciting science.

This book has developed from lecture courses and tutorials that we have taught at the Universities of Durham and Oxford. We are grateful to all the students who have given us much useful feedback. We would like to thank George Fleet for being an excellent mentor and Steve Davies for invaluable comments on the drafts of this book. BGD would also like to thank Karen.

Oxford
April 2002

B. G. D.
A. J. F.

Contents

1 Introduction

1.1 What are carbohydrates?

The term **carbohydrate** was originally derived in the nineteenth century from the French (hydrate de carbon) for the family of compounds possessing the empirical formula $C_n(H_2O)_n$. However, subsequently, the term has been greatly extended to encompass many other materials. In general, carbohydrates are polyhydroxylated aldehydes containing a number of carbon atoms, varying between 3 and 9. Some, such as glucose **1.1** (shown in the 6-ring cyclic form) should be very familiar. Others, such as sialic acid **1.2**, may look rather esoteric, but are in fact absolutely crucial throughout biology.

glucose
(β-glucopyranose)
1.1

sialic acid
1.2

1.2 Carbohydrates in nature

One thing is certain, that is, carbohydrates are everywhere, not just in the supermarket in 1 kg bags sold by Tate and Lyle. Glucose is the most abundant organic molecule on the planet. For example cellulose **1.3** is simply a polymer of glucose (actually β(1–4) linked); so next time you go outside and see a tree, remember that it is made of sugar. DNA and RNA are of course also made of sugar. Ribose **1.4** is a sugar containing 5 carbon atoms (called a pentose) that is a component of each of the building blocks used to make these two nucleic acids, which are fundamental to life itself.

ribose
1.4

cellulose = poly β(1–4)-glucopyranose
1.3

As we shall see later the prefix β(1–4) means that the 4 hydroxyl group of one sugar unit is linked to the 1 position (or anomeric centre) of the other, and that the stereochemistry of this linkage is equatorial (β), rather than axial (α).

Until relatively recently, the roles that sugars played in biological systems were thought to be limited to either acting as sources of energy, for example,

D-glucose in glycolysis to form ATP, or as good structural building blocks, for example, α(1–4) linked D-glucose as a polymer in starch, or as we mentioned β(1–4) linked D-glucose as a polymer in cellulose. However, it is now becoming ever clearer that carbohydrates are also key code molecules in biological communication events that control processes such as egg fertilisation, microbial infection, inflammation and cancer growth. The last chapter of this book illustrates the importance of these varied roles of sugars, and shows that considerations of sugar chemistry should always go hand-in-hand with a knowledge of their biological importance.

1.3   Misconceptions about carbohydrate chemistry

Students often find carbohydrate chemistry difficult for several reasons. The first is that sugars contain several stereogenic centres, and keeping track of all of this stereochemistry is difficult. This tends to put many students off straight away. However, what people forget is that most reactions involving potential changes in configuration occur at only one carbon atom, and therefore are in fact quite straightforward. Another problem that students encounter is just how to draw sugars themselves, again rather tricky considering all the stereochemistry that must be handled. For this reason we have added an Appendix to this book that has some guidelines (plus a few tricks) to help the reader do just this. Finally, the different potential modes of reactivity of free sugars, namely either as alcohols and/or as aldehydes, and also processes such as mutarotation and furanose/pyranose equilibration, can be rather confusing to the uninitiated. However, once the reader has grasped the fundamentals of acetal chemistry, which we cover several times in Chapters 2, 3, 5 and 6, the 'so-called' myths of sugar chemistry will be dispelled. They are simply remarkable molecules in which both alcohol and aldehyde functional groups are present.

2 Open chain and ring structure of monosaccharides

2.1 Aldoses and ketoses: open chain structures

Carbohydrates have the empirical formula CH_2O (hydrated carbon) and contain an oxygen atom attached to each carbon. Such compounds are therefore very highly functionalised. The commonest type of structures are the **aldoses** which have the general formula **2.1** and consist of a linear carbon chain with an aldehyde CHO group at C-1 (conventionally drawn at the top of a vertical chain), a varying number of carbon atoms which are secondary alcohols CHOH, and a primary alcohol at the other end of the chain. **Ketoses 2.2**, which have a primary alcohol at both ends and have a ketone within the chain, are less common. Most of this book will be about aldoses.

The simplest example of an aldose is **glyceraldehyde 2.3**, consisting of three carbons. Higher sugars containing four carbons **2.4** are known as **tetroses**, those with five carbons **2.5** as **pentoses**, and with six carbons **2.6** as **hexoses**. Glucose $C_6H_{12}O_6$ is an example of a hexose.

```
CHO              CH2OH
|                |
(CHOH)n          C=O
|                |
CH2OH            (CHOH)n
                 |
                 CH2OH

aldoses          ketoses
  2.1              2.2
```

```
CHO         CHO          CHO          CHO
|           |            |            |
CHOH        (CHOH)2      (CHOH)3      (CHOH)4
|           |            |            |
CH2OH       CH2OH        CH2OH        CH2OH

glyceraldehyde  tetrose     pentose      hexose
    2.3           2.4         2.5          2.6
```

Each of the carbon atoms that are secondary alcohols have four different groups attached and are **stereogenic (or chiral) centres**, giving rise to **stereoisomers**. To visualise their structures it is necessary to use conventions that allow three dimensions to be represented on paper in two dimensions. In the case of glyceraldehyde **2.3** in the margin, the stereoisomers will be **enantiomers** and are non-superimposable mirror images. Various alternatives are shown for glyceraldehyde with C-2 of glyceraldehyde in the plane of the paper; (i) with the carbon chain from left to right in the plain of the paper with the secondary hydroxyl group coming out of the paper for one enantiomer, and with the hydroxyl group behind the paper for the other enantiomer. Conventionally, as in (i), the hydrogen is usually not drawn. However, (ii) in which the hydrogen at C-2 is shown, is still a useful representation. The **Fischer convention** is commonly used for structures that have several stereogenic centres. Herein the stereogenic centres are represented as dots in the

D- or R-   L- or S-
glyceraldehyde
2.3

You should be familiar with the following stereochemical terms from a stereochemistry text book: enantiomer and diastereomer, conformation and configuration, staggered and eclipsed, relative and absolute configuration, D- and L- as descriptors of absolute configuration.
In addition, you should be able to apply the Cahn-Ingold-Prelog *RS* convention in order to define the configuration of a stereogenic (chiral) centre.

See Oxford Chemistry Primer No. 88, M.J.T. Robinson, *Organic Stereochemistry*, for further details.

The general descriptors *threo* and *erythro* are frequently used to indicate relative configurations of two adjacent stereogenic carbons in organic structures.

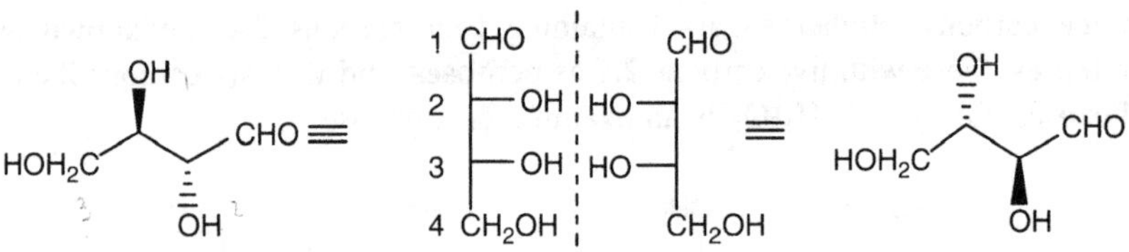

plain of the paper, the carbon chain is represented vertically by groups that go behind the plain of the paper, and the H and OH groups are drawn to the left and right, coming out of the paper as in (iii); the Fischer convention simplifies this by drawing all the lines as single lines as in (iv) and, usually, again leaves out the hydrogens. This is known as a **Fischer projection**. Thus the left column of structures represent the D-enantiomer of glyceraldehyde (in which the absolute configuration at C-2 is *R*) whereas the right-hand column shows different representations of the L-enantiomer (with an absolute configuration at C-2 of *S*).

2.1.1   Erythrose and threose

For sugars containing more than one stereogenic centre, stereoisomers will exist that are not related as object and non-superimposable mirror image, that is, compounds that are **diastereomers**. Generally for a sugar with *n* CHOH groups there will be 2^n stereoisomers. For higher sugars than glyceraldehyde (i.e. those containing more than three carbon atoms), the absolute configuration of the sugar is denoted by the prefix D- or L-, and depends on the absolute configuration of the secondary alcohol at the highest numbered stereogenic centre (this is *R*- for D-sugars and *S*- for L-sugars). The carbon atoms of aldoses are numbered so that the aldehyde terminus is always labelled as C-1 (carbon atom 1), and the other carbon atoms are numbered by counting along the chain away from the aldehyde as C-2, C-3 etc. There are four stereoisomers of the tetroses **2.4**: two diastereomeric pairs of enantiomers.

Stereoisomers that contain more than one stereogenic centre but only differ in configuration at *one* centre are called *epimers*.
Most naturally occurring sugars have a D-absolute configuration so that the configuration at the highest numbered carbon with a secondary alcohol—that is, that adjacent to the primary alcohol—has the same configuration as the stereogenic carbon in D-glyceraldehyde

Higher aldoses worth remembering the structures of are:
pentoses
ribose—component of RNA
hexoses
glucose
mannose—epimeric at C-2 with glucose
galactose—epimeric at C-4 with glucose

D-erythrose
2.7

L-erythrose
2.8

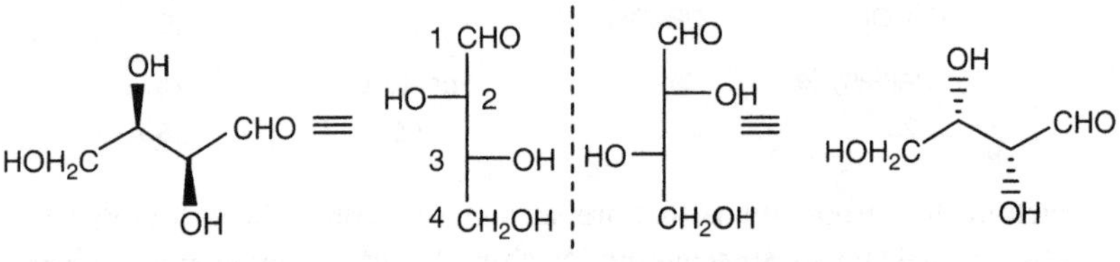

D-threose
2.9

L-threose
2.10

These two pairs of enantiomers are called **erythrose** (**2.7** and **2.8**) and **threose** (**2.9** and **2.10**). Each pair of enantiomers is diastereomeric with the other pair, that is, threose and erythrose are diastereomers. In this case the absolute configuration of each stereoisomer is determined by C-3 (not C-4) as this is the highest numbered stereogenic centre. The configuration is assigned as D or L according to whether the configuration at C-3 is the same as D- or L-glyceraldehyde. Deciding which is the case may most easily be achieved by looking at the Fischer projection with the aldehyde drawn at the top; if the hydroxyl group at C-3 is to the right (as it is for D-glyceraldehyde) then the sugar is D, if it is to the left (as for L-glyceraldehyde) then conversely it is L.

The zigzag drawing depicts a conformation of the carbon chain in which the substituents are *staggered* whereas the Fischer projection depicts a conformation in which the substituents are *eclipsed*—the least stable arrangement possible. Usually, in the organic chemistry literature, such structures would be drawn in their staggered conformations, but there are situations where Fischer formulae make it easier to see symmetry relationships; in this case, in particular, it is easier to assign the absolute configuration.

> It is worthwhile being able to handle both types of open chain structures and it is very important to be able to visualise these structures and to be confident of their equivalence – even for tetroses; both in regard to the these open chain structures and the cyclic structures that we shall come across shortly. Models make it very much easier to understand.

2.1.2 Pentoses and hexoses

Pentoses, with three stereogenic centres, exist as four diastereomeric pairs of enantiomers, and the hexoses as eight diastereomeric pairs. The D forms of these sugars are all drawn in the Fischer projections below. At this point it is worth drawing each of them in a zigzag formula and checking to see if you get them correct (i.e. whether they correspond to the structures shown at the end of this section).

The conversion of tetroses to pentoses (or of pentoses to hexoses) requires conversion of RCHO to RCH(OH)CHO with the introduction of a new stereogenic carbon centre. So, for example, ribose on extension of the carbon

> Synthetic methods for extending the carbon chain are discussed in Chapter 5.

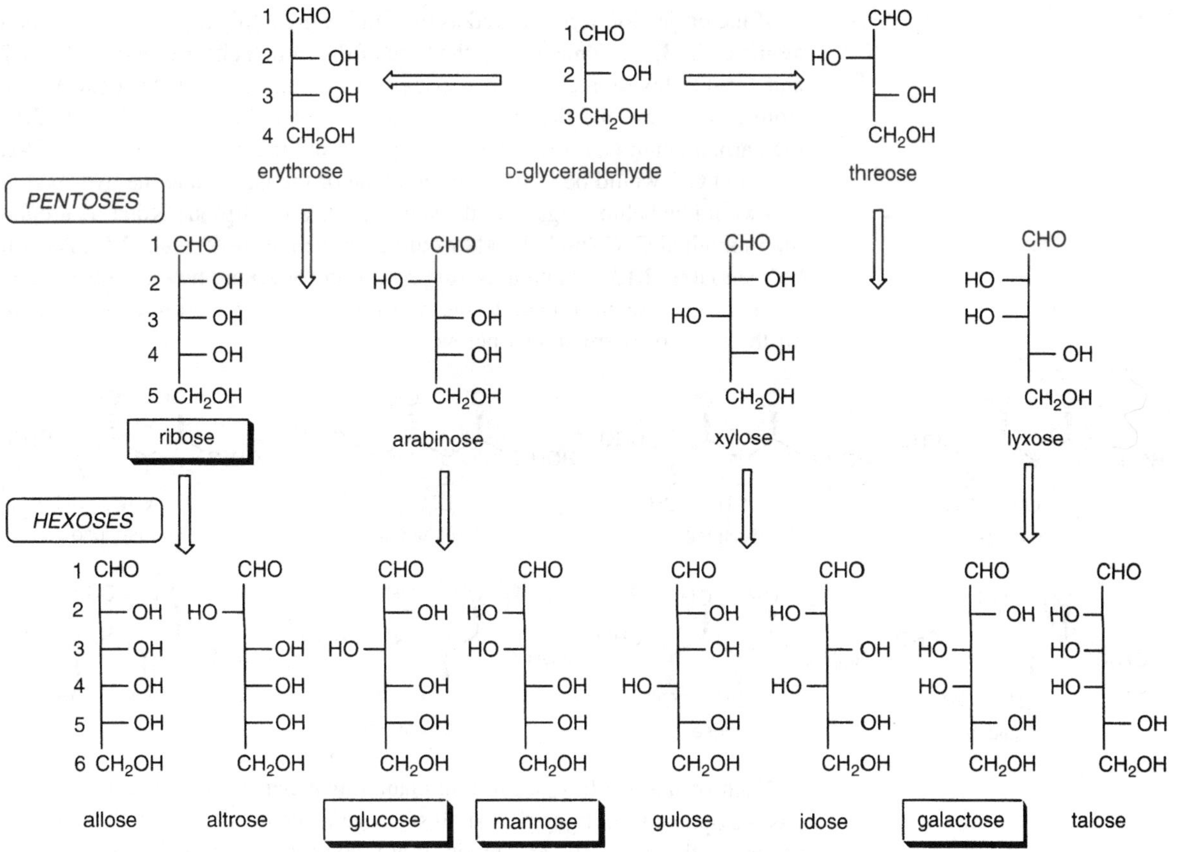

N.B. all structures are in the D-series

chain gives two new carbohydrates, allose and altrose, which only differ in configuration at the newly created stereogenic centre (C-2).

The open chain formula of D-glucose can be represented using the Fischer projection formula **2.11**. Each atom of the carbon chain when looked at is in the plane of the paper, with the two substituents to the left and right coming towards you. The adjacent carbons (above and below) in the chain are going behind the plane of the paper. This is a conformation in which all the substituents on adjacent carbons are eclipsed—that is, in the least stable possible conformation.

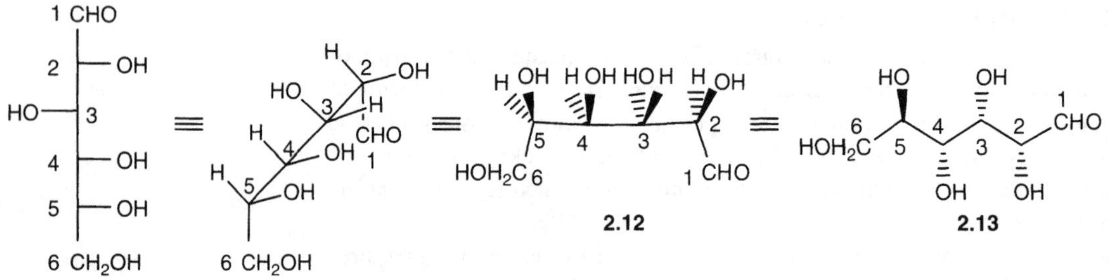

Models make it much easier to see the equivalence of these structures.

2.11
D-glucose

2.12

2.13

If the projection were viewed as though it were a 'pig trough' (with carbon atoms C-2, -3, -4, -5 making up the base of the trough channel and the C1–C2 and C5–C6 bonds making up the legs; for details see the Appendix), first from a side view and then from 90°, the view would be that of **2.12**. In **2.12** the carbon chain is in the plane of the paper and the hydroxyl groups at C-2, C-4, and C-5 would be in front of the plane of the paper, and the hydroxyl at C-3 would be behind (again all the substituents are eclipsed). Rotation around the individual C–C single bonds would give the zigzag formula **2.13**. As will be seen later, **2.13** is quite a useful projection for seeing how the open chain forms of glucose are related to the ring forms. The zigzag representations of all the D-hexoses are shown below:

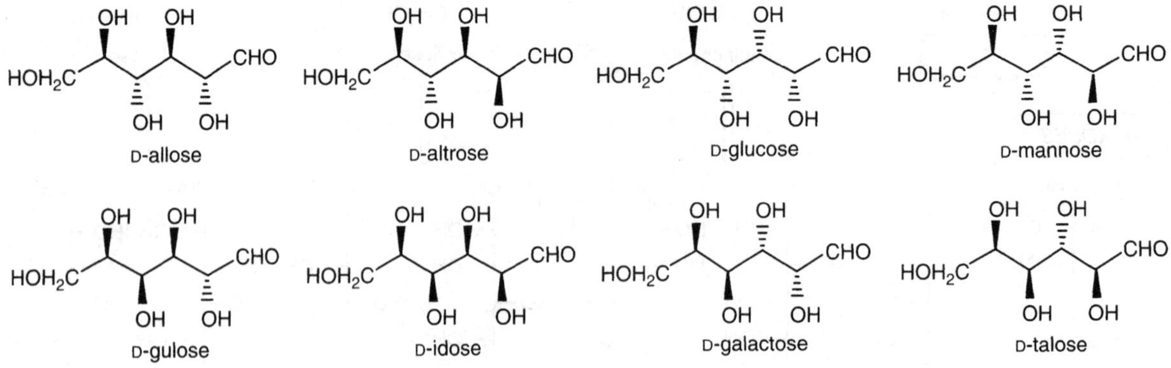

D-allose

D-altrose

D-glucose

D-mannose

D-gulose

D-idose

D-galactose

D-talose

Each of these D-hexoses has an enantiomer, namely the L-hexose, which has the opposite configuration at all stereogenic centres. Since in the majority of cases the D-sugars are much more common than their enantiomeric L-counterparts, apart from the case of some deoxy sugars (see below), the rest of this book will focus on the D-enantiomers of carbohydrates, and in particular the pentoses and hexoses. In fact when we omit the prefix D- or L- it will imply that we are talking specifically about the D-sugar. However, we must, of course, always remember the existence of the L-enantiomer!

2.1.3   Other monosaccharides

The most common monosaccharides are the D-aldoses that we have just met, which have oxygens attached to each of the carbons. The most commonly occurring ketose is D-fructose **2.14**, in which the configuration at its three stereogenic centres is the same as that at C-3, C-4, and C-5 of D-glucose **2.13** (and in fact mannose too). Deoxy sugars, which are usually more rare, lack an oxygen function at one or more of the carbon atoms. For example, 2-deoxy-D-ribose **2.15** is the sugar component of DNA. There are also some biologically important deoxy L-monosaccharides. L-Fucose **2.16** (6-deoxy-L-galactose) is involved in cell–cell recognition processes, and L-rhamnose **2.17** (6-deoxy-L-mannose) is a major component of plant cell walls.

D-fructose **2.14**   2-deoxy-D-ribose **2.15**   L-fucose **2.16**   L-rhamnose **2.17**

2.2   Ring structure of monosaccharides

2.2.1   Hemiacetals and lactols

Any given mixture of an aldehyde and an alcohol (e.g., methanol) is in equilibrium (catalysed by either acid or base) with a hemiacetal, a process which forms a new stereogenic centre. Since carbohydrates have both carbonyl and hydroxyl functions they are capable of forming intramolecular hemiacetals, known as **lactols**. In fact it turns out that the position of this equilibrium lies almost completely on the side of these cyclic hemiacetals, so that the amount of any open chain hydroxy-aldehyde present in either the solid form or a solution of a monosaccharide is negligible. To understand the chemistry of monosaccharides we need to consider both the kinetic and thermodynamic features that affect the rate, and position of equilibrium of this type of reaction.

2.2.2   Mechanism of the reactions

Hemiacetals are in equilibrium with their constituent aldehydes and alcohols by processes that are reversible in both acid and base. The reaction may be illustrated by the formation of a hemiacetal from an aldehyde with methanol. Thus in base catalysed hemiacetal formation the nucleophilic methoxide ion adds to the carbon–oxygen double bond to give a stable oxygen anion, **2.18**, which then picks up a proton from methanol to give the hemiacetal and regenerate the methoxide ion. In the base catalysed reverse process, that is, hemiacetal hydrolysis, a proton is removed from the hydroxyl group of the

hemiacetal

hemiacetal by the methoxide, to again give the tetrahedral anion, **2.18**, which then fragments into the aldehyde and the methoxide ion catalyst.

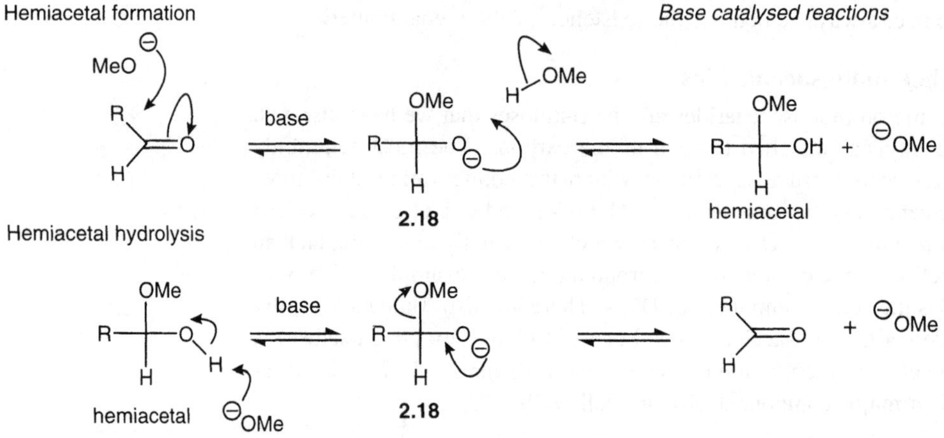

Hemiacetal formation

Base catalysed reactions

2.18

hemiacetal

Hemiacetal hydrolysis

hemiacetal

2.18

The most common significant stabilisation of a carbonium ion is by lone pairs of heteroatoms such as oxygen, nitrogen and sulfur; stabilisation of carbonium ions by adjacent oxygen lone pairs is one of the dominant features of carbohydrate chemistry.

In the acid catalysed pathway of hemiacetal formation, the carbonyl group of the aldehyde is first protonated to give oxonium ion **2.19**. This material can also be considered as an alternative resonance hybrid of a carbonium ion that is stabilised by the oxygen lone pair. Ion **2.19** (which is a much better electrophile than the neutral aldehyde) then undergoes nucleophilic attack by methanol (which is a weaker nucleophile than the methoxide ion of the base catalysed pathway) to give the protonated adduct **2.20**, which on loss of the proton catalyst gives the hemiacetal. For the reverse reaction, the breakdown of the hemiacetal is catalysed by addition of a proton to the oxygen of the hemiacetal bearing the alkyl group, to give the protonated adduct **2.20**. This is then followed by loss of methanol, with participation of the other oxygen lone pair, to give the oxonium ion **2.19** (or stabilised carbonium ion depending on your viewpoint) and then subsequent loss of a proton to give the aldehyde.

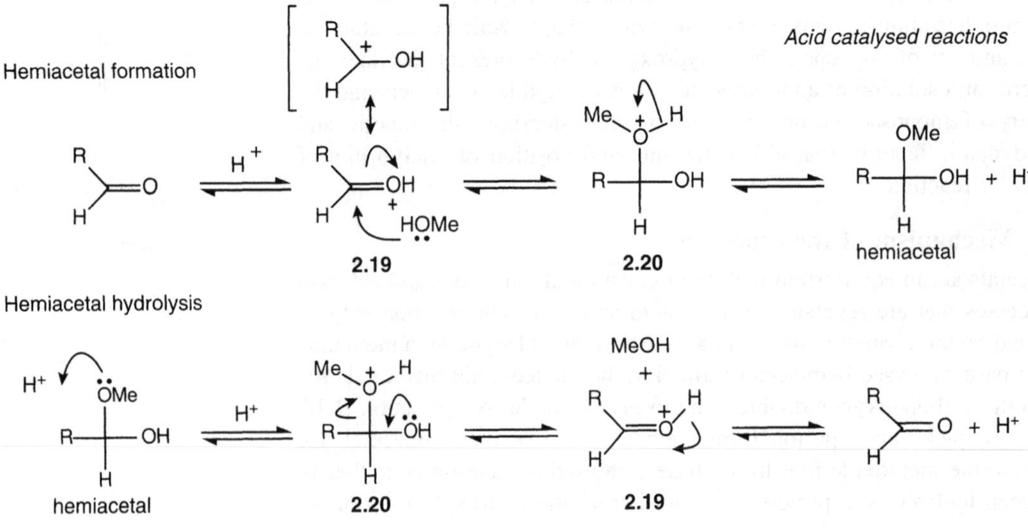

Hemiacetal formation

Acid catalysed reactions

2.19

2.20

hemiacetal

Hemiacetal hydrolysis

hemiacetal

2.20

2.19

2.2.3 Position of equilibria

Although the open chain representations of sugars that we have come across have some value, carbohydrates exist almost exclusively in a ring-closed form as a result of the cyclic version of hemiacetal formation, again catalysed by acid or base. Since sugars possess several hydroxyl groups, the formation of different ring size lactols (cyclic hemiacetals) is possible, most importantly containing 5- or 6-membered rings. To understand which forms are actually present we need to consider both kinetic and thermodynamic features of this chemistry.

First it is necessary to understand that, for a sugar possessing free hydroxyl groups, rather than forming a hemiacetal with an external alcohol in an intermolecular reaction, it is both kinetically and thermodynamically more favourable to form a cyclic hemiacetal (lactol) in an intramolecular process. This is because intramolecular reactions are favoured entropically over intermolecular ones.

Secondly it is important to remember that 5- and 6-membered rings are thermodynamically more stable than the corresponding 4- and 7-membered rings, since they are less strained. In particular 6-membered rings which can adopt a chair conformation are essentially free from all types of ring strain.

In cyclic frameworks 5- and 6-membered ring forming reactions using intramolecular hydroxyl groups are therefore favoured over other competing processes. Thus γ- and δ-hydroxyaldehydes are in equilibrium with 5-(furanose) and 6-(pyranose) ring forms:

For a detailed account of the thermodynamic stability of ring systems, including a full discussion of ring strain, and also an account of the kinetics of ring formation see Oxford Chemistry Primer No. 54, Martin Grossel, *Alicyclic Chemistry.*

Remember ΔS (the entropy change for the reaction) and $\Delta S^{\ddagger}$ (the entropy of activation for the reaction) contribute to ΔG (the Gibbs free energy change for the reaction) and $\Delta G^{\ddagger}$ (the activation energy of the reaction) as follows:

$$\Delta G = \Delta H - T\Delta S$$

$$\Delta G^{\ddagger} = \Delta H^{\ddagger} - T\Delta S^{\ddagger}$$

In general both ΔS and $\Delta S^{\ddagger}$ are more favourable for an intramolecular reaction.

5-ring lactol
furanose

6-ring lactol
pyranose

Let us now consider the situation for carbohydrates, which possess multiple hydroxyl groups, with particular reference to glucose. Thus the possible equilibria for glucose can be seen by first taking the flat projection formula **2.21**, and rotating about the C4–C5 bond so that the C-5 oxygen function comes into the plane of the paper and the CH$_2$OH group is sticking up out of the plane towards you, to give the flat projection **2.22**. It can then be seen that a 6-membered cyclic hemiacetal (or lactol) may be formed by nucleophilic attack of the C-5 hydroxyl group in **2.22** onto the carbonyl group of the aldehyde at C-1. As C-1 goes from being planar to tetrahedral, a new stereogenic centre is formed giving rise to two new diastereomers, **2.23** and **2.24**. Since the formation of hemiacetals is an equilibrium process, **2.23** and **2.24** are easily interconverted under acid or base catalysis via the open chain form **2.22**. These two diastereomers which arise from a change in the stereochemistry at C-1 (the anomeric position) are known as **anomers**. Diastereomer **2.23** in which the substituent at C-1 is down (or axial) is known as the α-anomer. Diastereomer **2.24** in which the C-1 substituent is up (or equatorial) is termed the β-anomer. This overall lactol ring forming process is sometimes known as the **cyclisation reaction**.

If you are not happy about the equivalence of the Fischer projection for glucose **2.11**, and the flat projection **2.21**, then see the Appendix.

The strict definition of α and β is whether the C-1 substituent is formally *cis* (α), or *trans* (β) to the oxygen atom at the highest numbered stereogenic centre, here C-5, when the sugar is drawn in the Fischer projection. However this precise definition is rather cumbersome, and for almost all purposes we can apply the simple rule of thumb that α is **axial** (down in D-sugars) and β is equatorial (up in D-sugars).

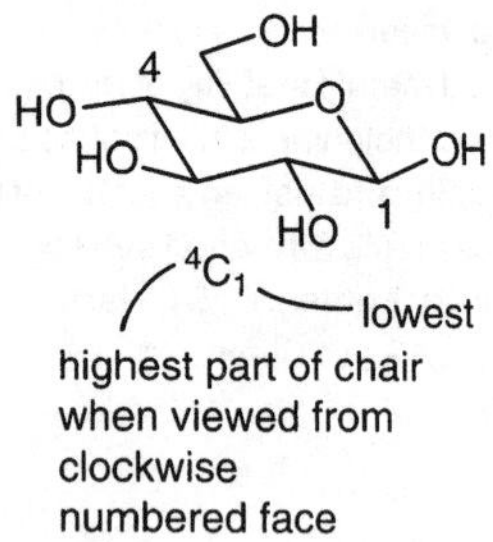

highest part of chair
when viewed from
clockwise
numbered face

The most stable conformation of
β-D-glucopyranose is known as the
$^4C_1$ conformation.

The 6-membered ring formed by reaction of the C-5 oxygen is known as the pyranose form of glucose (or glucopyranose). Although different conformations of the ring are clearly possible, a chair conformation is strongly preferred, as is generally the case for 6-membered rings. In particular for the β-anomer of glucose **2.26**, all the substituents in the $^4C_1$ chair form are in an equatorial arrangement. The only difference for the α-anomer **2.25** is that the hydroxyl group at the anomeric position is axial.

Easy equilibration of the lactols is observed in water and it can be either acid or base catalysed. In an equilibrium mixture in water there is 38% of α- and 62% of β-glucose, so the equatorial anomer is marginally more stable; we shall see in the next chapter that this is an exception to the normal observation that in the anomeric position, electronegative anomeric substituents are usually more stable in the *axial* configuration.

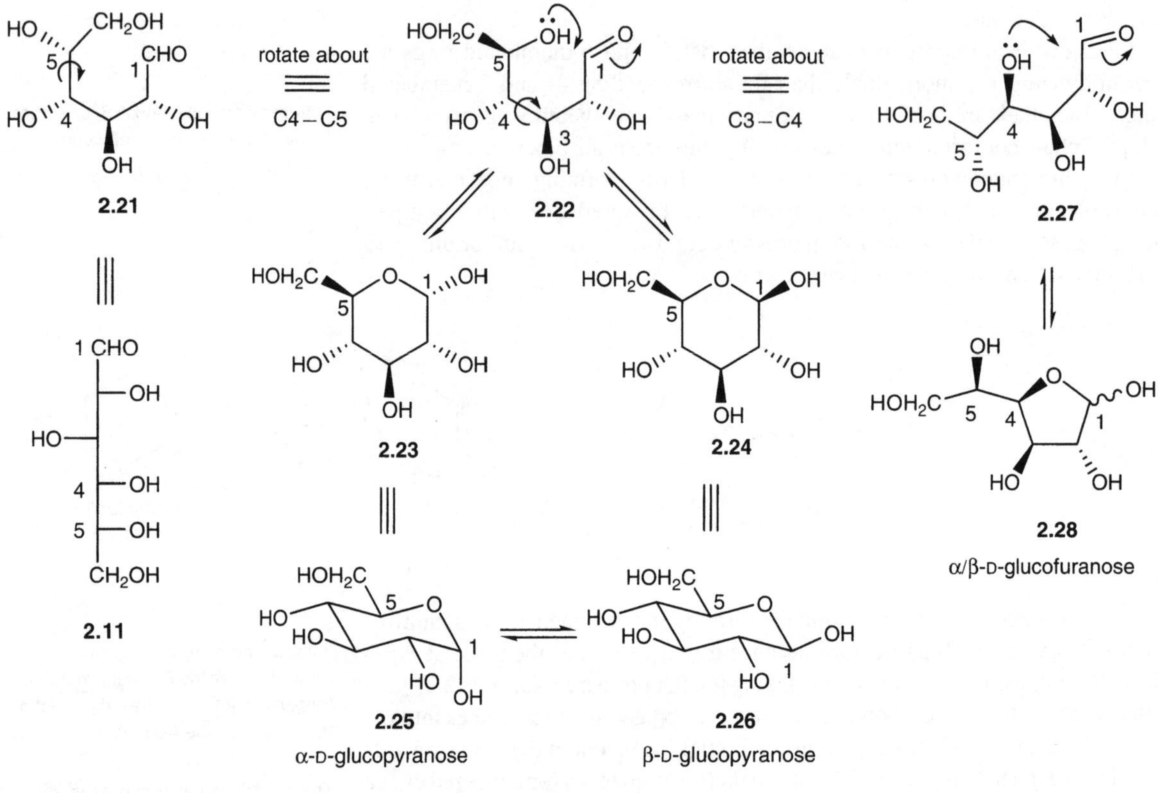

The wavy bond in **2.28** is a useful way
of representing mixed stereochemistry,
here showing the existence of both
α- and β-anomers.

There is also the possibility of a 5-ring form of the sugar being made by closure of the C-4 hydroxyl group in **2.27** onto C-1. This 5-membered ring form is known as the furanose form of glucose (or glucofuranose) and this closure would also give rise to α- and β-anomers, now of the furanose form **2.28**. However, in the case of glucose, the 6-membered ring form is the only form present in significant amounts under equilibrating conditions.

2.2.4   Ring structures of other carbohydrates

This book will mainly focus on the hexoses. Without exception, the pyranose structure is the preferred form adopted by the hexoses; they exist in aqueous solution as mixtures of α- and β-anomers, usually in the $^4C_1$ conformation detailed above for glucose. However, the sugars altrose, idose and talose, which would have several OH groups in axial positions in the $^4C_1$ conformation of the pyranose form, do have larger amounts of the furanose forms present at equilibrium (10–30%). The situation is similar for pentoses which also preferentially adopt the pyranose form, though ribose has about 20% of the furanose form at equilibrium. In the next chapter we shall focus in on the anomeric centre (C-1) and answer the questions: what in general are the relative proportions of the two α- and β-anomers present at equilibrium? and why?

2.3 Summary

At the end of this chapter you should be able:
- to draw the open chain structure of *aldoses* in both Fischer and zigzag projection formulae;
- to remember the structure of glucose, and hopefully that mannose is epimeric with glucose at C-2, and galactose is epimeric with glucose at C-4;
- to recall that carbohydrates usually exist in cyclic forms of hemiacetals (lactols) which are in equilibrium with each other by both acid and base catalysed routes. Both 5-membered rings (furanoses) and 6-membered rings (pyranoses) are common;
- to appreciate that diastereomers will arise due to the formation of a new stereogenic centre during the formation of the lactol from the open chain form and that these epimers are called anomers, which exist as α and β forms;
- to understand that the commonest conformation of pyranose rings of most sugars is $^4C_1$ and in β-D-glucopyranose all the alcohol substituents are in equatorial positions.

2.4 Questions

1. (a) Starting with the open-chain form of D-mannose draw out all the possible stable isomers that could be formed in aqueous solution.
   (b) Estimate the rough proportions of each.
2. (a) Draw other chair and boat conformations of D-glucose.
   (b) Label them ^xC_y and $^{x,y}B$ and $B_{x,y}$ accordingly.
   (c) Compare the number of axial substituents in the $^4C_1$ D-glucopyranose conformation with that in the $^1C_4$.
   (d) Of the D-hexoses, which are most likely to exist in roughly equal proportion of $^4C_1$ and $^1C_4$ conformations?

3 Reactions of the anomeric centre Part I

3.1 The anomeric centre

We saw in the last chapter that easy cyclisation of the open chain aldehyde form of sugars, which is catalysed by either acid or base, gives rise to cyclic hemiacetals or lactols. The resulting newly formed stereocentre at C-1 is known as the anomeric centre. For glucose in water, an equilibrium mixture of lactols consisting of 38% of the α-stereoisomer **3.1**, and 62% of the β-stereoisomer **3.2** is observed. So, in aqueous solution the equatorial anomer is marginally more stable than the axial one and only a very small amount of glucose exists in the furanose or open chain forms at equilibrium. We shall see shortly that this is an exception to the normal observation that at the anomeric centre, electronegative anomeric substituents are usually more stable in the *axial* orientation.

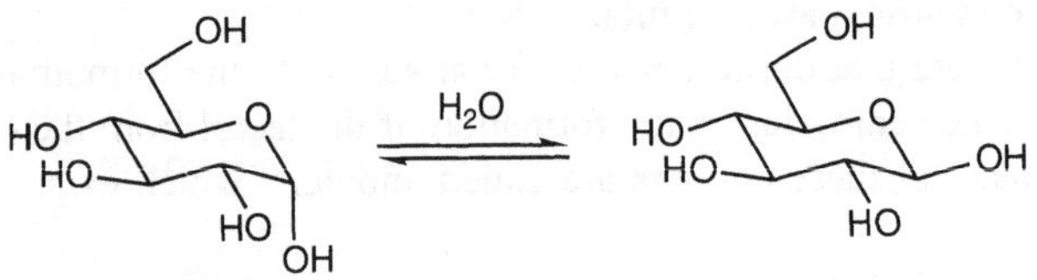

α-glucopyranose
38%
3.1

β-glucopyranose
62%
3.2

3.2 Acetal formation and hydrolysis

Acetals and hemiacetals are in the same oxidation state as carbonyls and it is therefore not surprising that they are readily inter-converted and you should think of the process of aldehyde-to-hemiacetal-to-acetal as one reversible reaction. In the case of sugars this is sometimes known as the cyclisation reaction.

Under acidic conditions hemiacetals undergo reaction with alcohols to form acetals. Although the mechanism for this reaction is similar to the one we encountered for hemiacetal formation in the last chapter, it is crucial to understand that acetal formation is *only* catalysed by acid and *not* by base. This is because the hydroxyl of the hemiacetal must first be protonated in order to convert the OH into a ready leaving group. Once protonation has occurred, the protonated hemiacetal can collapse to give an oxonium ion, such as **3.3**. This oxonium ion is then nucleophilically attacked by the alcohol, in this case methanol, to yield the acetal product. Overall, this is an S_N1 process.

Acetal formation

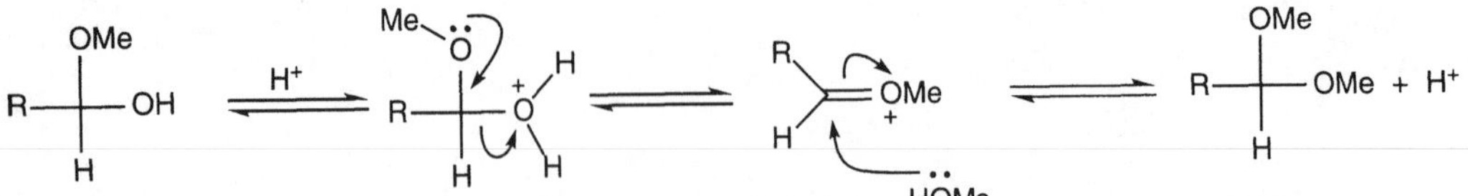

hemiacetal

3.3

acetal

Again this is an equilibrium reaction and so acetals can be readily hydrolysed by treatment with aqueous acid. Once again this reverse reaction is *only* catalysed by acid and *not* by base, and again passes through an oxonium ion intermediate in an S_N1-type pathway. However, for the hydrolysis reaction the nucleophile that attacks the oxonium ion **3.3** is water, and therefore leads to the formation of the hemiacetal.

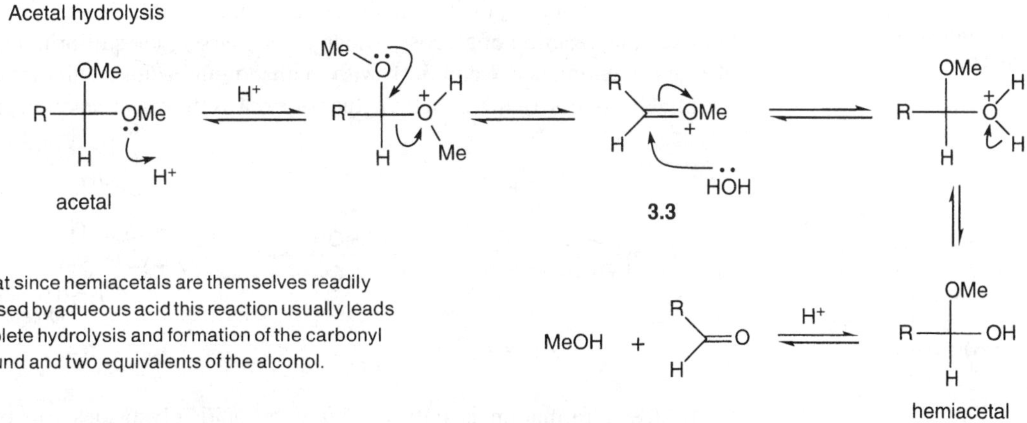

Acetal hydrolysis

Note that since hemiacetals are themselves readily hydrolysed by aqueous acid this reaction usually leads to complete hydrolysis and formation of the carbonyl compound and two equivalents of the alcohol.

hemiacetal

3.3 Fischer glycosidation

Lactols, such as the pyranose forms of glucose **3.1** and **3.2** are hemiacetals and, as such, are able to react with alcohols to form acetals, that is, structures with two single C–O ether bonds attached to the same carbon atom. The mechanism of this type of acetal formation is the same as the one outlined above, the only difference being that the hemiacetal is cyclic. As for all hemiacetals, it is the presence of an oxygen substituent α to the anomeric carbon that makes any carbocation formed there more stable than it would otherwise be. This S_N1 process is therefore relatively easy under acid catalysis, and in the case of carbohydrate lactols the reaction initially gives rise to the formation of a cyclic carbocation intermediate called a glycosyl cation. This glycosyl cation is then attacked by the alcohol in the second step, to yield the acetal product.

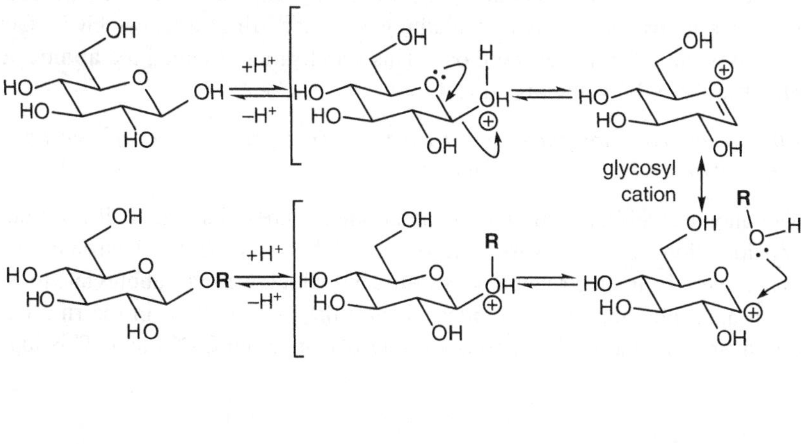

glycosyl cation

Thus, in the presence of an acid catalyst, an alcohol can react with a sugar lactol hemiacetal to form an acetal, which is called a **glycoside**. This process is known as a **Fischer glycosidation** reaction. Of course, this is an equilibrium reaction and so treatment of any glycoside with aqueous acid can reverse this formation to give the parent carbohydrate. Usually a source of anhydrous Bronsted acid (e.g. $SOCl_2$ mixed with MeOH) or a Lewis acid (e.g. $ZnCl_2$) in the presence of excess alcohol is used to drive the equilibrium to completion. In addition the reaction is invariably under thermodynamic control, and, therefore, the thermodynamically most stable product is formed. For example, reaction of glucose (which is, of course, the equilibrium mixture of α and β anomers **3.1** and **3.2**), with methanol and acid, results in the preferential formation of the α-methyl glucoside **3.4** as the major reaction product.

3.1 / 3.2   3.4

As acetal formation is only catalysed by acid, glycosides can only be hydrolysed by the reverse reaction initiated by treatment with aqueous acid, and are stable under almost all other reaction conditions. This means that the anomeric centre can effectively be protected as a glycoside at the beginning of any reaction sequence, simply by treatment of the free sugar with an alcohol and acid.

3.4   The anomeric effect

The above observation that the Fischer glycosidation reaction of glucose with methanol and acid produces mainly the α-glucoside is at first perplexing. As we have already said, this type of acid catalysed acetal formation is under thermodynamic control, and therefore the product we expect to observe is the thermodynamically most stable one. However, any simple consideration of steric effects would imply that the β-stereoisomer, with the anomeric substituent in the equatorial position, should be more stable than the α counterpart, where this substituent is in a more hindered axial environment. The isolation of the α compound as the thermodynamically preferred product indicates to us that there is something else occurring here, which in fact stabilises the α form over the β one. This stabilisation is called the **anomeric effect** and can be summarised as follows:

Electronegative substituents on a pyranose ring prefer to occupy an axial rather than an equatorial orientation

The anomeric effect is in fact a general phenomenon, but we shall limit our consideration here to sugars. The origin of the effect is an electronic one. When there is an electronegative substituent, which we shall abbreviate to X, in an axial position at the anomeric centre the oxygen atom in the ring has one of its lone pairs of electrons anti-periplanar to the C–X bond. This lone pair can participate in a stabilising two electron interaction, which can be formally represented by the two resonance forms **3.5** and **3.6**. In molecular orbital terms this corresponds to an $n \rightarrow \sigma^*$ interaction, namely partial donation of the oxygen lone pair (in an *n* orbital) into the anti-bonding (σ^*) orbital of the C–X bond. The net effect of this electronic interaction is electronic delocalisation, and therefore it is stabilising.

Note that the n and σ* orbitals are syn-periplanar; ideal for overlap.

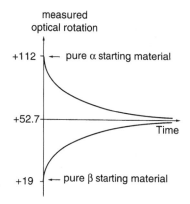

3.5   **3.6** $n \longrightarrow \sigma^*$
donation

Resonance form **3.6** will only be a contributor if X is happy with a formal negative charge; this, therefore, explains why the anomeric effect only operates when the anomeric substituent is an electronegative atom, such as O, F, Cl, or Br. If the anomeric substituent X is in an equatorial position it is not possible for the ring oxygen to orientate a lone pair anti-periplanar to the C–X bond, and hence there can be no such stabilising electronic interaction.

In molecular orbital energetic terms only σ* orbitals associated with atoms of sufficient electronegativity are low enough in energy to interact with the lower lying n lone pair orbital.

There are several consequences of the anomeric effect; for example a shortening of the O–C ring bond length, and corresponding lengthening of the C–X bond, which can in fact be measured. However, we only really need to concentrate on the overall general conclusion, which is that electronegative substituents at the anomeric centre prefer to be axial. This has many implications throughout carbohydrate chemistry, not least of which is the fact that α-glycosides tend to be thermodynamically more favourable than β ones.

3.5 Mutarotation

Having now considered the anomeric effect let us return to the question of the preferred form of glucose in aqueous solution, which we encountered in Section 3.1. If, for example, we take a sample of pure α-glucose **3.1**, dissolve it in water, and measure the optical rotation of the sample we see that this decreases from an initial value of +112, to an equilibrium value of +52.7 over time. Likewise, if we perform the same experiment with pure β-glucose **3.2**, the optical rotation increases over time from an initial value of +19, to this equilibrium value of +52.7. This change in optical rotation with time is termed **mutarotation**, and results from equilibration of the α and β forms. There are two things we need to be concerned about here. First, what is the mechanism of this equilibration reaction?

The first step towards equilibration occurs by opening of either of the pyranose forms of glucose **3.1** and **3.2** to give the open chain aldehyde **3.7**. This reaction is hemiacetal hydrolysis, and as we saw in Chapter 2 it is subject to acid and base catalysis. Open chain aldehyde **3.7** may then re-cyclise by one of several ring-closing processes, to reform a hemiacetal.

measured
optical rotation

+112 ← pure α starting material

+52.7

Time

+19 ← pure β starting material

These ring closing reactions are hemiacetal formation reactions and are, therefore, also catalysed by acid or base. The first possibility is regeneration of the 6-membered pyranose ring by attack of the 5-hydroxyl group back onto the C-1 aldehyde. However, since this hydroxyl group may attack *either face* of the aldehyde, either the α-pyranose (**3.1**) or the β-pyranose (**3.2**) products *may* be formed. In this way the α and β forms, **3.1** and **3.2**, may equilibrate via the open chain aldehyde, and this equilibration reaction is catalysed by either acid or base.

Remember 6-membered rings are usually more stable than 5-membered ones, but 5-membered rings are usually formed faster.

However, we should also be mindful that the open chain hydroxy aldehyde, **3.7**, may also ring close by nucleophilic attack of the free 4-hydroxyl group onto the aldehyde, thus forming a 5-membered ring. Again this hydroxyl group may also attack either face of the aldehyde, leading to either the α (**3.8**) or β (**3.9**) furanose products. All the above compounds **3.1**, **3.2**, **3.7**, **3.8**, **3.9** are therefore in equilibrium with each other.

This brings us to our second consideration, namely the position of equilibrium. As we saw in the last chapter glucose exists almost entirely in the pyranose form, and only extremely small amounts of material are present as either the furanose or open chain forms. This is a consequence of the thermodynamic stability of the 6-membered ring, particularly in the case of glucose where all the substituents can adopt equatorial positions. As noted previously other sugars, for example ribose, altrose, idose and talose, have much larger amounts of the furanose form present at equilibrium.

Finally, we return to consider the relative amounts of the α and β pyranose products **3.1** and **3.2**. In light of the anomeric effect we may initially have expected the α form **3.1** to completely dominate the equilibrium. However, we must be mindful that on steric grounds equatorial substituents are favoured over axial ones, and so steric effects may, at least in part, counteract the anomeric effect. There is therefore, competition between electronic effects, which favour the α anomer, and steric effects which may favour the β anomer. The fact that the β anomer is the major form of glucose present at equilibrium in aqueous solution is usually ascribed to solvation. One rationale is that in aqueous solution the anomeric hydroxyl group is heavily solvated, thus increasing the effective steric bulk of the anomeric substituent. The net result is that steric effects win out, and there is an overall slight preference for the equatorial β anomer. The situation is however reversed in the case of mannose, which is epimeric at C-2 to glucose. In this case the α anomer dominates at equilibrium, in line with the anomeric effect.

3.6 Summary

At the end of this chapter you should be able:
- to understand the chemical and structural relationship between carbonyls, hemiacetals and acetals;
- to work through and recall the mechanisms of acetal formation and hydrolysis and apply them to the Fischer glycosidation reaction;
- to recall that as a result of the *anomeric effect* electronegative anomeric substituents are more stable in an axial orientation;
- to understand the underlying reasons for the anomeric effect;
- to be aware of some of the other consequences of the anomeric effect;
- to understand and describe the process of mutarotation, and why it is subject to general acid/base catalysis;
- to be able to give mechanisms for the inter-conversion of α and β anomers and furanose and pyranose forms of sugars via open-chain aldehydes;
- to be able to rationalise the relative proportions of α/β and furanose/ pyranose forms present at equilibrium for glucose in aqueous solution.

3.7 Questions

1. $SOCl_2$ or CH_3COCl in MeOH are good sources of dry acid in alcohol. By drawing out the mechanism of the reaction of these compounds with MeOH work out why.
2. Explain why the position of the equilibrium shown below is solvent dependent.

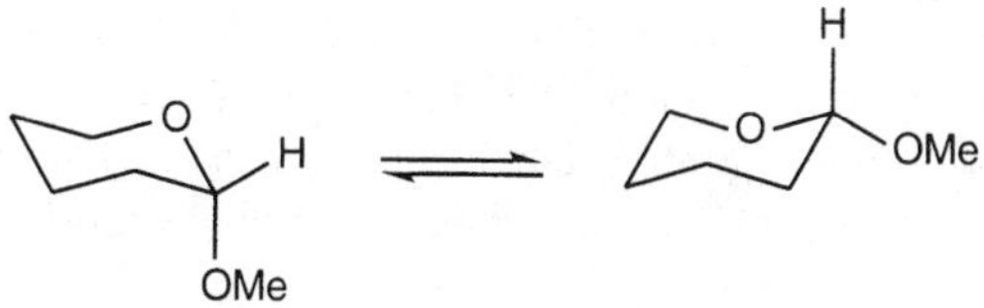

In CCl_4	83	:	17	
In H_2O	52	:	48	

3. Deduce the structure of the product of the following reaction, and give a mechanism for its formation.

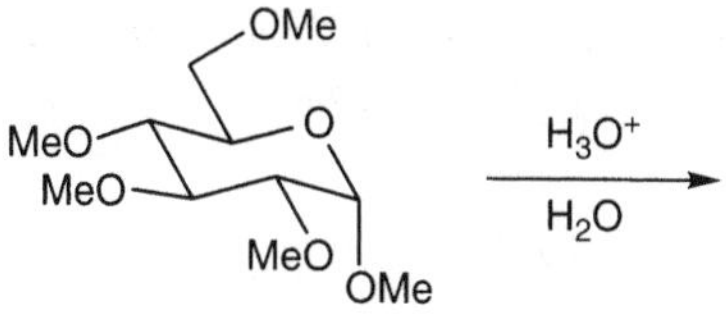

4. Explain why equilibration of the following two compounds by acid or base in $H_2^{18}O$ is 30 times faster than incorporation of ^{18}O from the solvent.

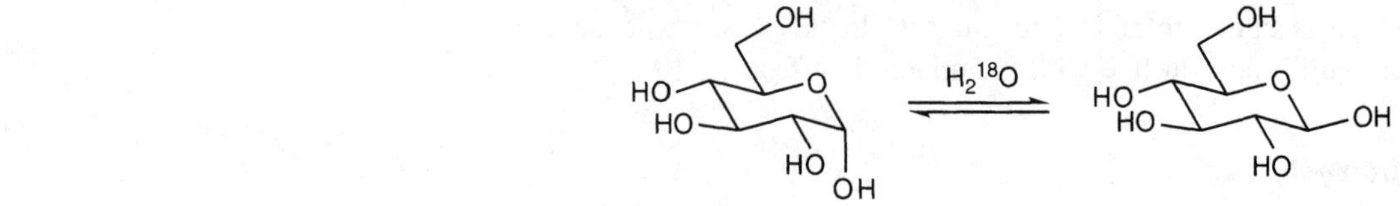

5. Deduce the structure of the major product that is formed during the following reaction and give a mechanism for its formation.

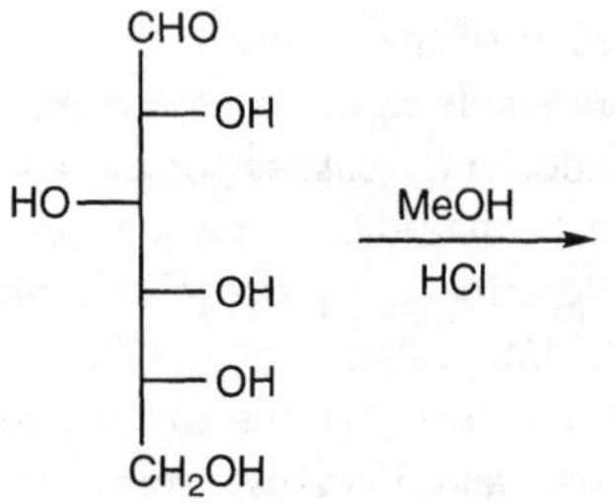

6. Explain the following observations. Reaction of compound **I** with *p*-toluenesulfonic acid in methanol at 25 °C for 24 h gave a single product **II**, which was stable under the reaction conditions for up to 5 days. Treatment of **II** with the same reagents at reflux gave a product mixture consisting of **III** (55%) and **II** (45%).

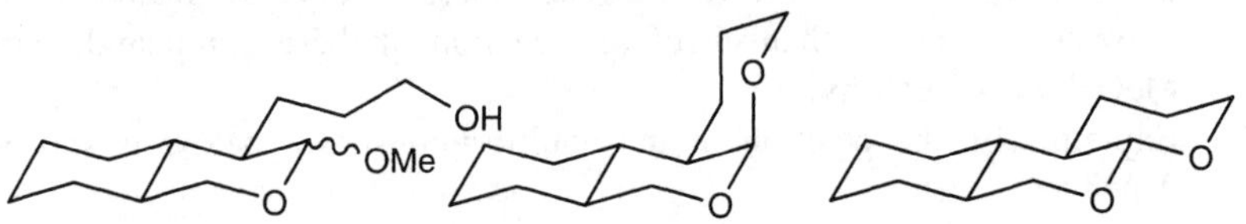

4 Reactions of hydroxyl groups Part I

This chapter will focus on the hydroxyl groups of sugars and investigate their characteristic modes of reactivity. As we shall see, there is nothing particularly new or mysterious about this sort of Chemistry, the complications arise only because of the fact that there are so many of them! We shall simply encounter reactions in which the sugar hydroxyls react typically as nucleophiles directly analogously to non-sugar alcohols, and, in fact, the majority of the Chemistry, namely the formation of esters and ethers, should already be familiar to you. Discussion of formation of cyclic acetals by reaction of diols is discussed separately in Chapter 6.

There are essentially three types of alcohol groups in carbohydrates:

(i) the primary hydroxyl group ($-CH_2OH$), which for D-glucose is at C-6

(ii) the secondary hydroxyl groups ($-CH(OH)-$) which in D-glucopyranose are at C-2, 3, and 4

(iii) the anomeric hydroxyl group which is at C-1.

primary OH-6   secondary OH-2,3,4   anomeric OH-1

In Chapter 5 we shall focus on reactions that are specific to the anomeric centre itself and much of that chemistry may be less familiar. However, one of the major complications of sugar chemistry, and in fact one of the reasons that many students find it baffling, is the fact that the functionality at the anomeric centre can react in two ways—as an alcohol or as an aldehyde—and we must always bear this possibility in mind. Most commonly the anomeric centre can be tied up as a *glycoside*, by itself acting as an aldehyde to form an acetal and this effectively protects the anomeric centre and prevents any further reactivity. We saw in the previous chapter that the formation of methyl glycosides is particularly straightforward and that these materials are stable under most sorts of reaction conditions, barring acid. However, if we are dealing with a free hydroxyl at the anomeric position, for example the cyclic hemiacetal (or lactol), **4.1**, then the situation becomes more complicated. On the one hand, it is clear that this hydroxyl can react as a nucleophile, in parallel with the other sugar hydroxyls. Therefore,

The following hydroxyl group derivatives are discussed in this chapter:

Carboxylate Esters
Acetates $-O(CO)CH_3$ or $-OAc$
Benzoates $-O(CO)C_6H_5$ or $-OBz$

Sulfonate Esters
Tosylate $-O(SO_2)C_6H_4CH_3$ or $-OTs$
Mesylate $-O(SO_2)CH_3$ or $-OMs$
Triflate $-O(SO_2)CF_3$ or $-OTf$

Ethers
Benzyl $-OCH_2C_6H_5$ or $-OBn$
Trityl $-OC(C_6H_5)_3$ or $-OTr$
Trimethylsilyl
    $-OSi(CH_3)_3$ or $-OTMS$
Tert-butyldimethylsilyl
    $-OSi(CH_3)_2(C(CH_3)_3)$ or $-OTBS$
    or $-OSiMe_2Bu^t$ or $-OTBDMS$
Tert-butyldiphenylsilyl
    $-OSi(C_6H_5)_2(C(CH_3)_3)$ or
    $-OSiPh_2Bu^t$ or $-OTBDPS$

A methyl glycoside

Remember a glycoside is a general term for a sugar that possesses a substituent at the anomeric position, which is not a free hydroxyl group.

we would expect the cyclic form **4.1** to display reactivity typical of an alcohol and it is this type of reaction that we shall consider in this chapter. However, as we saw in the previous chapter, under appropriate conditions the cyclic hemiacetal **4.1** is in fact in equilibrium with the open chain hydroxy aldehyde forms **4.2** and **4.3**. These open chain forms **4.2** and **4.3** can, of course, react as aldehydes, and reactions of this type will be discussed further in Chapter 5.

4.1§	**4.2**¶	**4.3**¶	**4.4**§
α-anomer			β-anomer

§In this form the anomeric centre acts as an alcohol
¶In this form the anomeric centre acts as an aldehyde

Another complication is that (as we also saw in Chapter 3) the process of *mutarotation* allows equilibration of α- and β-hydroxyls at the anomeric position, that is, the equilibration of the α-anomer **4.1** with the β-anomer **4.4** via these open chain forms. Therefore when considering reactivity of the anomeric centre as a hydroxyl group we also have to bear in mind potential competing inter-conversion between these α and β forms. Finally, of course, this equilibrium also allows inter-conversion between the pyranose (6-ring) and furanose (5-ring) forms of the sugar. Although much of the discussion in this book will focus on the reactions of hexopyranoses (and almost invariably pyranose forms of the D-hexoses dominate), we should always bear in mind that conversion to a furanose form is a possibility under certain reaction conditions, and, in fact, we shall come across reactions that proceed via furanose forms in Chapter 6.

One of the first things that the synthetic organic chemist has to consider when attempting chemistry with sugars is solubility. Generally speaking, the presence of several free hydroxyl groups means that these so-called 'free sugars' are very polar in nature. Consequently, they are very soluble in polar solvents, especially those that are capable of H-bonding to these free hydroxyls (e.g. water) but completely insoluble in the standard, more non-polar solvents used for performing organic reactions, product purification and manipulation (e.g. dichloromethane, ether, ethyl acetate, toluene etc.). Furthermore, the selective protection of particular hydroxyl groups allows the regioselective reaction of those left unprotected; something that is often crucial in, for example, chemical oligosaccharide synthesis (Chapter 7). Therefore, the most common starting point of any reaction sequence involving sugars is to protect the majority, or even all, of the free hydroxyl groups. We shall now look at several ways of achieving this.

4.1 Acetylation reactions

Alcohols react readily with activated carboxylic acid derivatives such as acid anhydrides, or acid chlorides under appropriate conditions to produce the corresponding esters. Since esters are essentially non-nucleophilic and stable to a wide range of reaction conditions they are frequently used as protecting groups. In addition esters are in general much less polar than the corresponding alcohols.

Acetates are the most commonly used esters since all but the most hindered tertiary hydroxyl groups can be readily esterified simply by stirring the alcohol in a mixture of acetic anhydride and a base. This base not only serves to mop up the equivalent of acetic acid that is produced during the reaction but it also catalyses the reaction itself, be it either by general base or nucleophilic catalysis. This is necessary since the uncatalysed reaction of alcohols with acetic anhydride is very slow at room temperature. An alternative method is to use Lewis acid catalysis for the acetylation reaction. However since mutarotation at the anomeric centre is catalysed by both bases and Lewis acids then it is necessary to bear in mind the possibility of this competing process. In fact acetylation of free sugars is usually undertaken in one of three ways, each of which has different stereochemical consequences for the anomeric position.

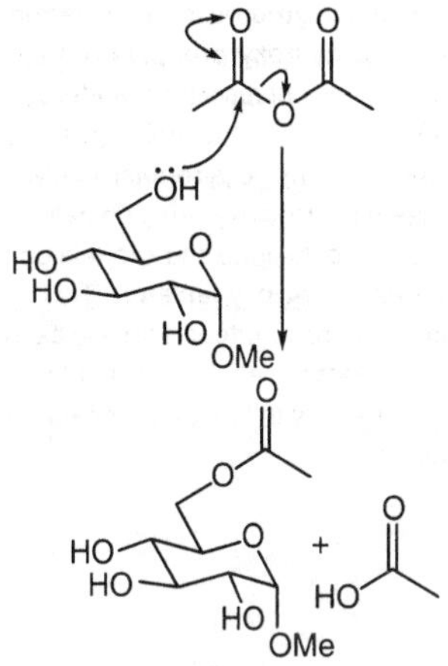

Addition/elimination mechanism of ester formation

4.1.1 Acetylation with acetic anhydride/pyridine

Free sugars come as an equilibrium mixture of stereoisomers at the anomeric centre, or **anomers**. Thus glucose in a bottle is a mixture of the α and β forms **4.1** and **4.4**. The different proportions of the α and β (and also pyranose and furanose) forms depends on the particular sugar in question. Acetylation of glucose with acetic anhydride and pyridine, often used in equal proportions as the reaction solvent, at room temperature produces the fully acetylated compounds **4.5** and **4.6** as a mixture of anomers, which corresponds directly to the composition of the anomeric mixture in the free sugar starting material. This result is a consequence of the fact that under these reaction conditions the acetylation of the free hydroxyl at the anomeric centre is faster than the competing process of opening of the hemiacetal and mutarotation.

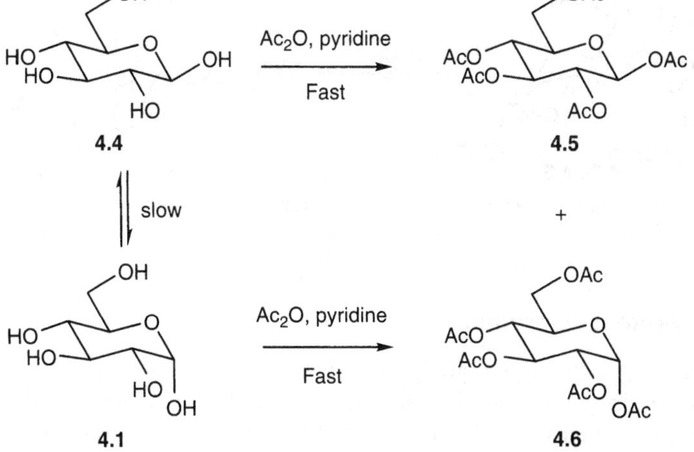

Here acetylation is faster than α/β equilibration via the open chain form. Therefore a mixture of products is formed. The relative proportions of the α- and β-anomers in the product mixture corresponds with the α/β mixture of anomers in the starting material.

A β-hydroxyl group is more nucleophilic than the α-hydroxyl group, not only because it occupies a less hindered equatorial, rather than axial, position, but also because of a repulsive effect between the ring oxygen lone pair orbitals and the lone pair orbitals on the β-anomeric oxygen atom. This stereoelectronic effect is known as the *kinetic anomeric effect* and should not be confused with the anomeric effect itself.

4.1.2   Acetylation with acetic anhydride/sodium acetate

Sodium acetate is a relatively weak base. As such, acetylation with sodium acetate and acetic anhydride is sluggish at room temperature. Therefore acetylation is usually carried out at, or near, reflux (~100 °C) and the fully acetylated products of these reactions are usually the β acetates, as, for example, in the case of glucose compound **4.5**. This selective formation of the β product is rationalised by the fact that at these elevated temperatures the mutarotation, which leads to α/β equilibration, is much faster than the actual acetylation reaction itself. Since the equatorial β hydroxyl in **4.4** is more nucleophilic than the axial α counterpart in **4.1**, the β component reacts in preference. The net overall result is selective formation of the β-pentaacetate as the major reaction product.

4.1	**4.4**	**4.5**
α-anomer	β-anomer	β product formed preferentially

4.1.3   Acetylation with acetic anhydride/Lewis acid catalysis

Acetylation can also be catalysed by the addition of a Lewis acid such as zinc chloride to the reaction mixture. However, strong Lewis acids catalyse equilibration of the final product α- and β-acetates. Therefore equilibration of these α- and β-acetates will occur under the reaction conditions. Since the α-anomer is thermodynamically favoured by the anomeric effect, the predominant reaction product is the α-pentaacetate, as, for example, in the case of glucose α-pentaacetate **4.6**.

4.1 + 4.4	**4.5 + 4.6**	**4.6**
Mixture of anomers	Mixture of anomeric acetates	Equilibrated to form mainly the α product

For a detailed account of the use of protecting groups in general see Oxford Chemistry Primer No. 95, Jeremy Robertson, *Protecting Group Chemistry.*

4.2   Protecting groups

Protecting groups should satisfy several important criteria. First, they should be formed in good yield, they should be stable to subsequent reaction conditions, and, finally, they should be readily removed under appropriate conditions. Their importance is demonstrated by the chemical synthesis of disaccharides which will be discussed in Chapter 7 and relies on selective access to particular hydroxyl groups of sugars; this can usually only be achieved by protection of the others. It is, therefore, not surprising that much of the chemistry of monosaccharides is concerned with protection. We shall now survey briefly some of the most commonly encountered hydroxyl protecting groups as applied to carbohydrates and in particular highlight some selective protection processes.

Since the primary hydroxyl group of a carbohydrate is in a sterically less encumbered environment than the other hydroxyls it is therefore more nucleophilic. Consequently, selective reaction of this hydroxyl group in the presence of other free hydroxyls is possible. This is particularly the case when we attempt reactions with bulky reagents. This is in fact an underlying theme of the reactions of sugar hydroxyl groups and we shall see several examples of selective reaction of the primary hydroxyl group in the rest of this chapter.

4.3 Acetals

The use of cyclic acetals as protecting groups for both sugar hydroxyls and the anomeric centre will be discussed at length in Chapter 6. However, we should just note in passing the fact that protection of the anomeric position as a glycoside as discussed in Chapter 3 is taken as read in most synthetic sequences involving carbohydrates. The reader may like to take note of the fact that the anomeric centre is tied up in this way in almost all of the remaining reactions discussed in this chapter.

4.4 Ether protecting groups

4.4.1 Benzyl ethers and variants

Benzyl ethers (ROBn) are one of the most common protecting groups used in sugar chemistry. Benzyl ethers may be made simply by treatment of an alcohol, for example, **4.7** with benzyl halides, most commonly benzyl bromide, in the presence of a good base such as sodium hydride. The result is that all free hydroxyl groups are benzylated to yield **4.8**. The mechanism of this reaction is a simple S_N2 displacement of the halide anion by the alkoxide which itself is formed by deprotonation of the alcohol. This reaction can be catalysed through the addition of a source of iodide ion (e.g. tetra-butylammonium iodide); iodide is both a good nucleophile and leaving group. The iodide therefore displaces the bromide, before being displaced itself—this is known as nucleophilic catalysis.

4.7   NaH, BnBr / DMF   **4.8**

Benzyl ethers may also be formed under mild conditions of acid catalysis using benzyl trichloroacetimidate with, for example, trifluoromethanesulfonic

4.7 CF_3SO_3H **4.8**

via

para-methoxybenzyl (PMB)

DDQ

etc. etc.

Resonance stabilisation
of the trityl cation

acid (triflic acid, CF_3SO_3H). This again leads to the formation of **4.8** from **4.7**.

Benzyl ethers are most readily cleaved under the very mild conditions of catalytic hydrogenation, most commonly with a heterogeneous catalyst such as palladium on carbon or palladium black. Palladium hydroxide (Pearlman's catalyst) is also a very effective catalyst, particularly for the removal of sterically inaccessible/multiple benzyl groups.

The *para*-methoxybenzyl (PMB) ether has also found increasing use in carbohydrate chemistry over recent years since it may be selectively removed in the presence of many other protecting groups. Formation is achieved as for benzyl ethers via the use of *para*-methoxybenzyl chloride. PMB ethers may be cleaved either by catalytic hydrogenation as for benzyl ethers, or in fact selectively by oxidation with one electron oxidising agents such as ceric ammonium nitrate (CAN), or 2,3-dichloro-5,6-dicyanobenzo-1,4-quinone (DDQ).

4.4.2   Trityl ethers

Trityl (triphenylmethyl) ethers are protecting groups that are useful in being very selective for the primary hydroxyl group. Formation occurs via an S_N1 type process and is achieved by treatment of the unprotected sugar, for example the galactose derivative **4.9**, with trityl chloride in pyridine. Due to the very bulky nature of the reagent only the primary hydroxyl group will react, here producing **4.10**. This leaves other secondary hydroxyl groups free for further reaction, for example for protection with an orthogonal protecting group, such as benzyl, to produce the fully protected compound **4.11**. The trityl ether may be cleaved under mildly acidic conditions to regenerate the primary hydroxyl group. Therefore a multiple step sequence involving protection of the primary hydroxyl, benzylation of the remaining free hydroxyls, and de-tritylation allows access to monosaccharides possessing only a primary free hydroxyl group, such as **4.12**.

4.9 $\xrightarrow[\text{pyridine}]{Ph_3CCl}$ **4.10** $\xrightarrow[\text{BnBr}]{NaH}$ **4.11** $\xrightarrow{AcOH}$ **4.12**

4.4.3   Silyl ethers

Silyl ethers are one of the most commonly used alcohol protecting groups. As such a full discussion of their uses is not possible in this limited amount of space and the reader is referred to the sources listed in the Further Reading for a more comprehensive discussion. It should suffice to say that silyl ethers are readily formed by treatment of the appropriate sugar, for example, the mannose derivative, **4.13**, with the silyl chloride (or occasionally the silyl triflate) in the presence of a weak base such as imidazole. Thus, unhindered pertrimethylsilyl (TMS) ethers of carbohydrates such as **4.14** may be readily formed by treatment with trimethylsilyl chloride (TMSCl). By the use of

'ate'
complex

a bulky silylating reagent, such as *tert*-butyldiphenylsilyl chloride (TBDPSCl) or even *tert*-butyldimethylsilyl chloride (TBDMSCl or TBSCl) good regioselectivity may be obtained—as expected selective reaction of the less hindered primary hydroxyl may be achieved, for example, to yield **4.15**.

Imidazole acts as a nucleophilic catalyst and forms intermediates such as:

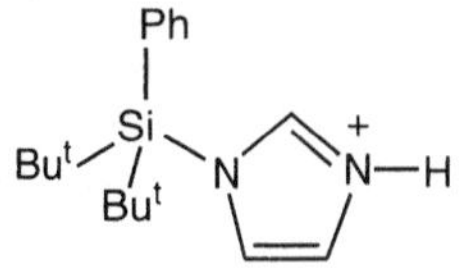

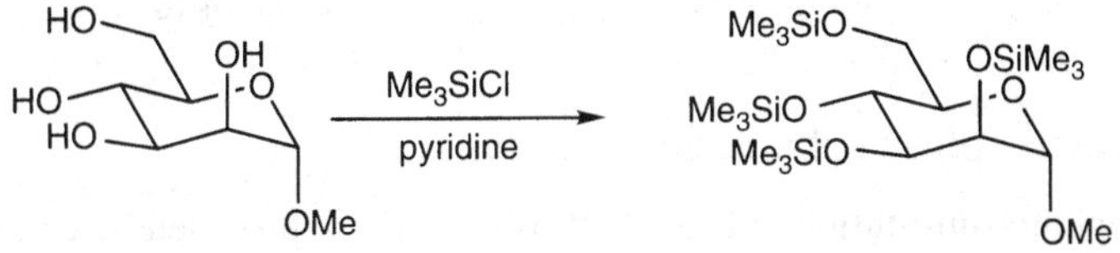

The relative rates of acid hydrolysis of silyl ethers show how they increase in stability in the order:

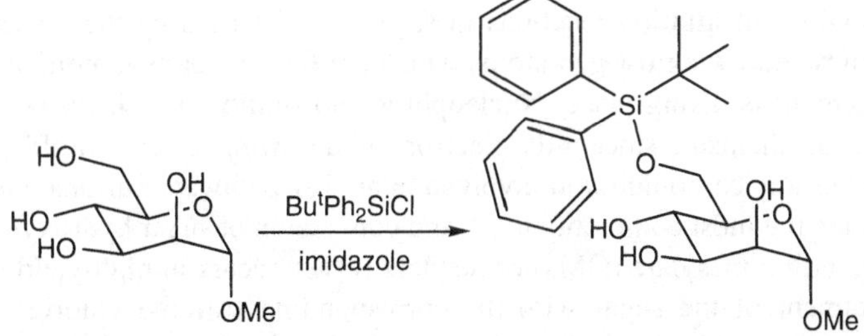

R−OSiMe$_3$ 1
R−OSiMe$_2$Bu^t 1×10^{-3}
R−OSiPh$_2$Bu^t 1×10^{-5}

Silyl ethers may be removed either by treatment with acid, which may also cause removal of other acid labile protecting groups such as acetals and trityl ethers, or more selectively by treatment with a source of organic soluble fluoride, such as tetrabutylammonium fluoride (TBAF).

4.5 Ester protecting groups

As we saw in Section 4.1 acetates derived from sugar hydroxyl groups are readily formed by treatment with acetic anhydride in pyridine. Benzoyl esters (ROBz) are also very commonly used ester protecting groups and are readily formed by treatment of the alcohol with benzoyl chloride in the presence of an amine base, such as triethylamine.

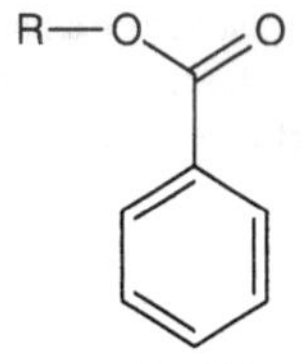

Both these types of ester protecting groups are readily removed by treatment with a suitable nucleophile, most commonly methoxide, in a trans-esterification reaction. The most frequently used conditions are either simple treatment with potassium carbonate in methanol, or sodium methoxide in methanol (Zémplen deacetylation). In both these cases methoxide is, in fact, catalytic. Milder reaction conditions, such as treatment with primary amines, are also effective for the removal of certain esters and can be used for selective deprotection reactions in particular circumstances.

= R−OBz
benzoyl ester

4.5.1 Sulfonate esters

Sulfonate esters may also be used as protecting groups and have the added ability to act as good leaving groups for nucleophilic substitution reactions. The toluenesulfonate (tosylate) ester is widely used as it is more stable than other sulfonate esters, such as trifluoromethanesuflonate (triflate, CF$_3$SO$_3$−) or methanesulfonate (mesylate, CH$_3$SO$_3$−) whilst still reactive enough to be displaced subsequently. Tosylate esters of the OH-6 group such as **4.16** may be selectively formed using tosyl chloride in pyridine.

4.13 → **4.16**

pyridine, 0°C

4.6 Nucleophilic substitution reactions

mesylate   triflate

tosylate

Nucleophilic displacement of the sulfonate esters of secondary hydroxyl groups is inhibited in carbohydrates by the β-*oxygen effect*. An approaching nucleophile experiences electrostatic repulsion from the lone pair electron density of neighbouring oxygen atoms. This repulsion is greater for charged nucleophiles. For this reason the primary hydroxyl group of a sugar is easier to displace since it has only one oxygen substituent in a β-position whereas OH-2 has three.

Nucleophilic substitution of sugar hydroxyls must be preceded by conversion of the OH into a leaving group. It should be borne in mind that in fact nucleophilic substitution reactions in sugars are retarded by the presence of electron withdrawing oxygen atoms which are β to the carbon atom at which displacement is taking place. Nucleophilic substitution invariably occurs by an S_N2 mechanism, since this electron withdrawing effect would greatly destabilise any carbonium ion involved in an S_N1 pathway. Sulfonate leaving groups are the most commonly used, and conversion of sugar hydroxyls to the corresponding mesylate (OMs) or tosylate (OTs) occurs in high yield simply by treatment of the sugar with the corresponding sulfonyl chloride in the presence of a mild base such as pyridine. In fact, as we saw above, in many cases, selective tosylation of primary hydroxyl groups can be achieved under appropriate conditions. Triflate (trifluoromethanesulfonate, OTf) is in fact an even better leaving group, and triflates are simply formed by treatment of the alcohol with trifluoromethanesulfonic anhydride (triflic anhydride) and pyridine. These materials are unstable and so are often used directly for subsequent nucleophilic displacement reactions.

Nitrogen is most readily introduced into a carbohydrate via nucleophilic displacement of a leaving group with azide. These displacement reactions are S_N2 reactions which proceed with inversion of configuration. Subsequent reduction of the azide to an amine allows access to amino sugars.

Other good nucleophiles such as iodide will also displace leaving groups via S_N2 reactions with inversion of configuration. Iodine may also be introduced by treatment of an unprotected carbohydrate with mild iodinating agents based on triphenylphosphine and iodine/carbon tetraiodide, or variants thereof. Such reactions are often selective for the replacement of only the primary hydroxyl group leading to primary iodides such as **4.17**.

Epoxide formation may be achieved by intramolecular nucleophilic displacement of a leaving group by a free hydroxyl on the neighbouring carbon atom. Since this reaction occurs via an S_N2 process, this hydroxyl group must be necessarily *trans* to the leaving group. Thus, treatment of tosylate **4.18** with base results in formation of the epoxide **4.19**.

Epoxide opening with nucleophiles occurs regioselectively under stereoelectronic control to preferentially produce the diaxial product. Therefore, epoxide **4.19** undergoes attack by added nucleophiles specifically at C-2, resulting in the formation of products **4.20**. The reaction is stereospecific in that epoxide opening proceeds exclusively with inversion of configuration to produce the *trans* diaxial product. Typical nucleophiles include azide and thiols, though epoxide opening may also be achieved by reducing agents such as $LiAlH_4$, to allow the formation of deoxy sugars.

4.18 **4.19** **4.20**

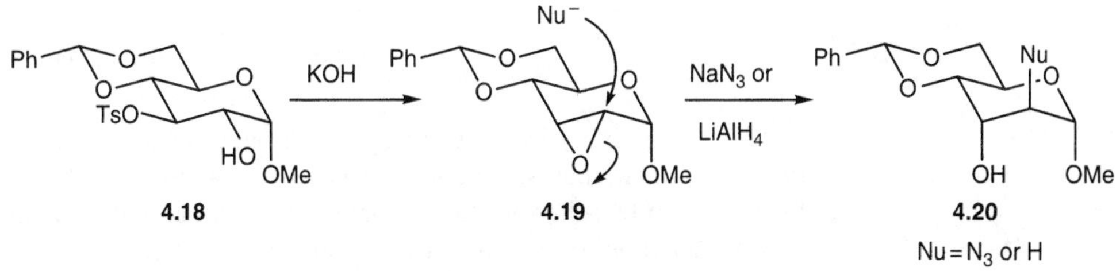

4.7 Oxidation

Treatment of sugars possessing a single free secondary hydroxyl group, of which diacetone glucose (**4.21**) is a prime example, with oxidising agents such as pyridinium chlorochromate (PCC) allows the formation of ketones. Similar oxidation reactions are readily achieved with the usual plethora of oxidising agents, for example, Swern, Pfitzner-Moffat, etc. These ketones may then undergo the usual range of reactions. For example, Wittig type reaction followed by hydroboration, which occurs on the least hindered face of the molecule, allows the formation of carbohydrates with branched carbon chains such as **4.22**. In fact, simple reduction of the so formed ketones can also be useful, since, often, this allows access to sugars which are epimeric to

You should make sure you know the mechanism of chromate oxidations and the common mechanism to the Moffatt-type oxidations.

We shall see how to synthesise diacetone glucose **4.21** in Chapter 6.

Epimers are diastereomers whose configuration is inverted at one stereogenic centre.

4.21 diacetone glucose

4.23 diacetone allose

4.22

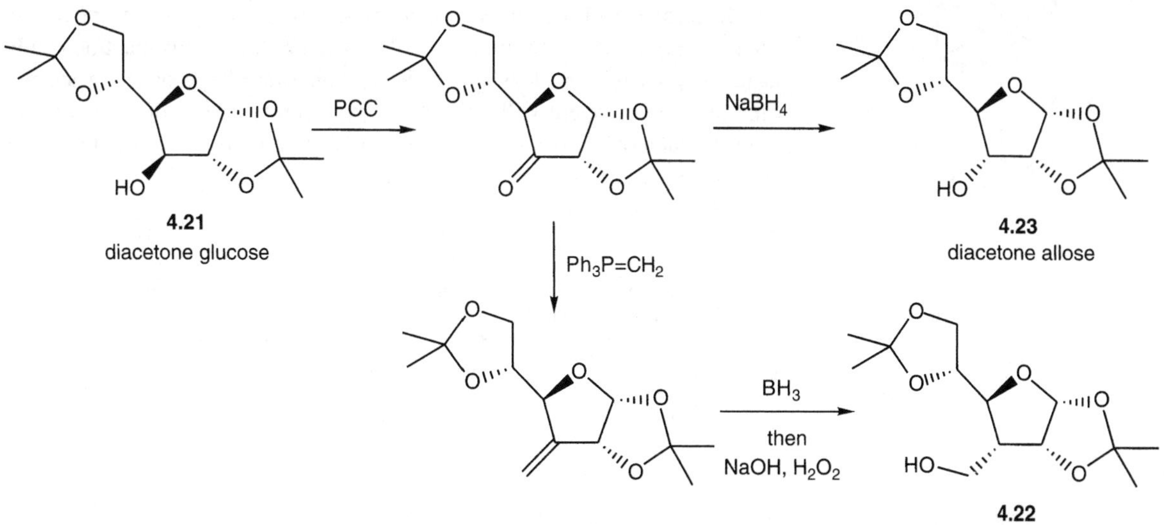

the starting material. The case of diacetone glucose is particularly useful; reduction of the ketone occurs from the less hindered face of the molecule, anti to the bulky acetonide protecting groups, allowing access to the *allo* product **4.23**.

Sugars possessing a free primary hydroxyl group may be oxidised either to the aldehyde, or all the way to the carboxylic acid, which is called a *uronic acid*, depending on the potency of the oxidising agent used.

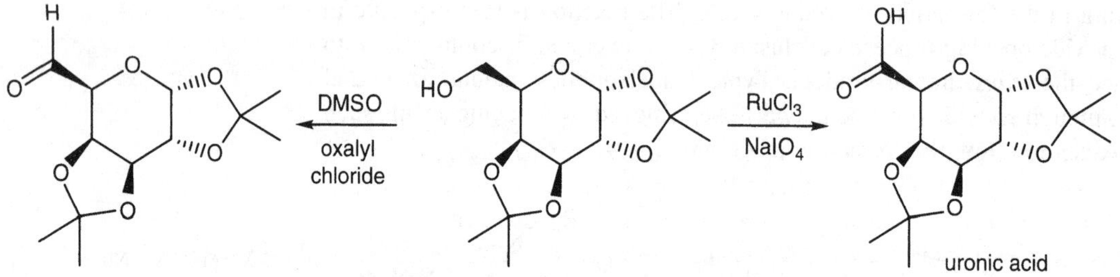

Usually the other hydroxyl groups must be protected to prevent competitive oxidation, but sometimes selective oxidation of only the primary hydroxyl can be achieved, for example, by oxidation with oxygen and Pt in aqueous solution to yield the uronic acid **4.24** directly.

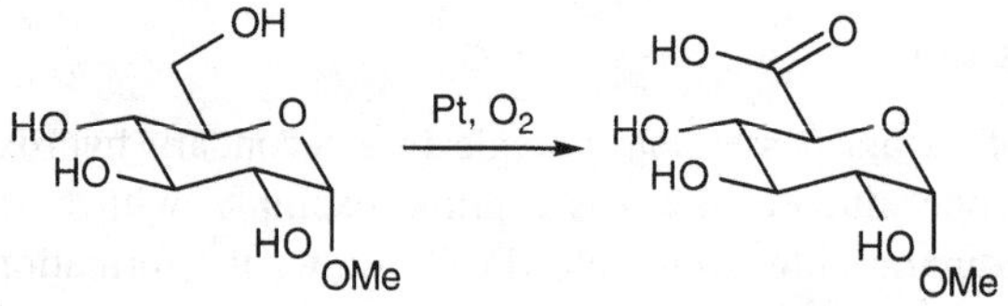

4.24

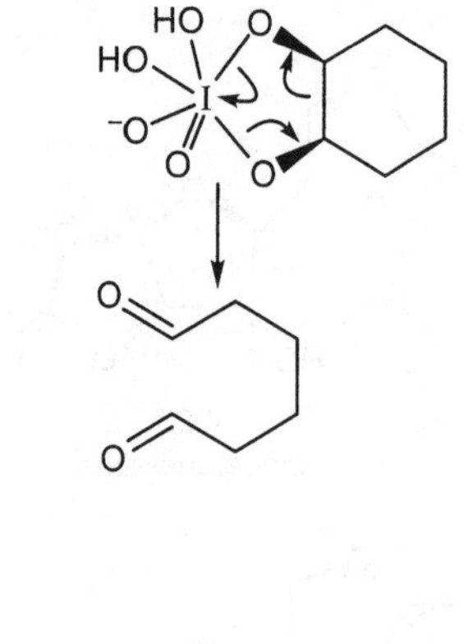

Nitric acid is powerful enough to oxidise both the primary hydroxyl group and the anomeric aldehyde at C-1 to give open chain polyhydroxy dicarboxylic acids called **aldaric acids**. Other methods that allow the selective oxidation of the anomeric centre are discussed in Chapter 5.

Finally, we should, in passing, mention another type of oxidation reaction. Sodium periodate is a commonly used reagent in organic synthesis for the specific cleavage of 1,2 diols and, as expected, periodate cleavage of carbohydrate *cis* 1,2 diols produces two aldehyde products. This reaction can be particularly useful for cleavage of the carbohydrate backbone at a specific point, since the reagent will only cleave between two alcohol groups which are *cis* to each other. Alternatively, it may also be used for shortening of the carbohydrate chain by one carbon atom.

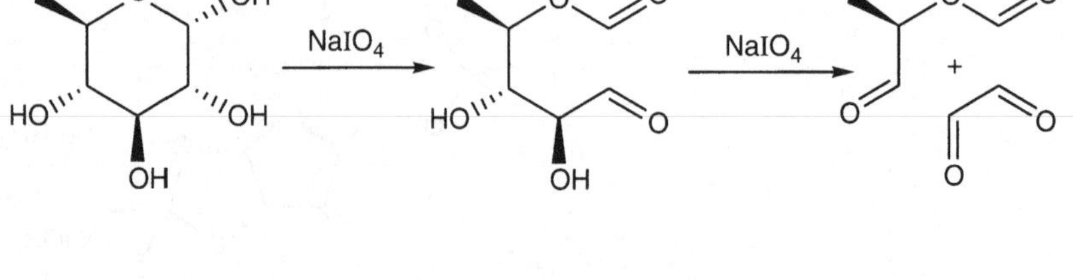

4.8 Reduction

Deoxy sugars constitute an important class of materials, not least due to their existence as sub-units of many important biologically active natural products. Therefore several efficient methods for the removal of hydroxyl groups have been developed.

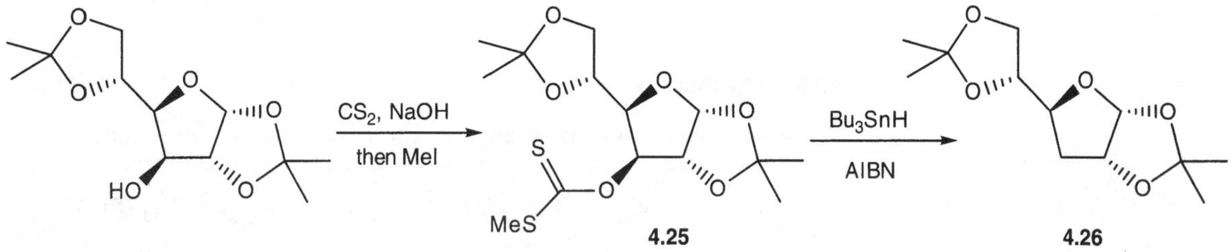

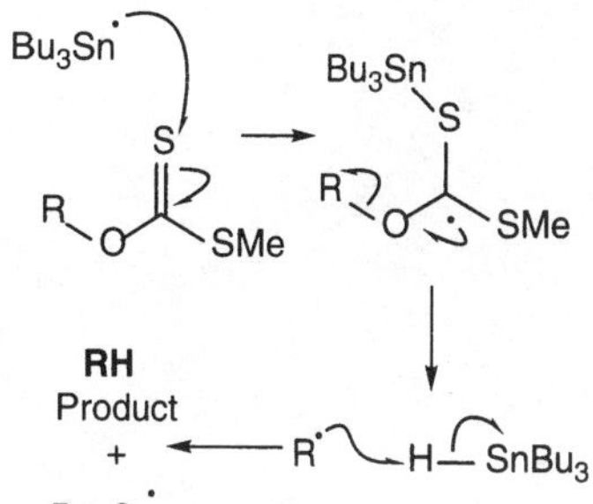

4.25   **4.26**

One of the most useful of these is the Barton–McCombie deoxygenation, which proceeds via a free radical process. Thus formation of the xanthate ester (e.g. **4.25**), by reaction of the alcohol with carbon disulfide and subsequent methylation, is followed by treatment with tributyltin hydride/AIBN to produce the deoxygenated material, **4.26**, in good yield.

It should be noted that nucleophilic displacement of leaving groups, for example OMs or OTs, with hydride from reducing agents such as $LiAlH_4$ is not generally a good route to deoxy sugars. Since, as we previously mentioned, nucleophilic substitution is much slower in carbohydrates due to the β-oxygen effect, then under the basic conditions produced by these reducing agents side reactions such as elimination reactions occur preferentially. Hydrogenolysis of halides in the presence of a mild base, for example, triethylamine, leads to the replacement of halide by hydride and also provides a good method of creating deoxy sugars. Reductions, specifically at the anomeric centre, will be discussed in Chapter 5.

4.9 Summary

At the end of this chapter you should be able:

- to describe the three types of hydroxyl groups found in sugars;
- to remember that carbohydrates exist in cyclic forms of hemiacetals [lactols], in which the anomeric centre acts as an alcohol, which are in equilibrium with each other [mutarotation] and with open chain forms in which the anomeric centre acts as an aldehyde;
- to realise the importance of reacting these hydroxyl groups [protection] to allow regioselective reactions of those left unprotected;
- to recall and explain the products of acetylations under different conditions;
- to give examples of methods that are used to selectively protect the primary hydroxyl group of a sugar;
- to recall the methods, with mechanisms, for forming and cleaving ether [benzyl, trityl, silyl] and ester [acetate, benzoate, sulfonate] protecting groups;
- to understand the difficulties associated with the substitution of sugar hydroxyl groups;

- to recall the usefulness of sulfonate esters as powerful leaving groups;
- to describe methods for making azide and halide substituted sugars;
- to recall methods for the oxidation of non-anomeric hydroxyl groups give ketones, aldehydes, carboxylic acid derivatives, including the ox dative cleavage of diols;
- to recall methods and mechanisms for the reduction of non-anomer hydroxyl groups to give deoxy sugars.

4.10   Questions

1. How would you convert glucose into the following compounds?

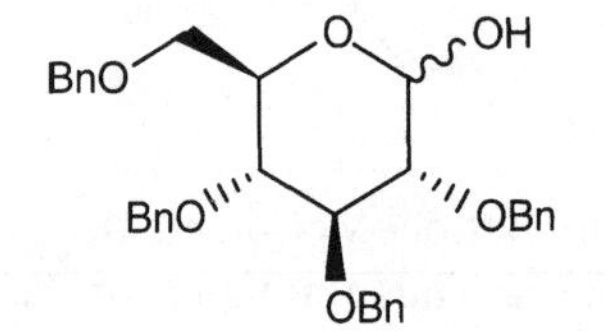

(i) 2,3,4,6 tetra-*O*-benzyl glucose 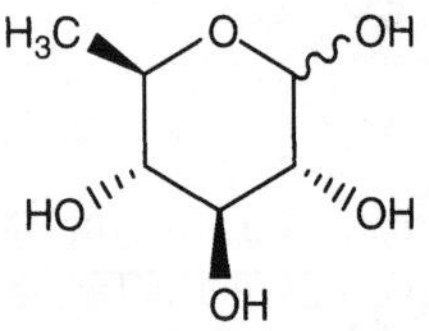   (ii) 6-deoxyglucose

2. (i) Explain why the two epoxides **A** and **B** both react with sodiu hydroxide in ethanol to give the same diol.
    (ii) Epoxides **A** and **B** also react with sodium azide in DMF to each gi a single azide of molecular formula $C_{14}H_{17}N_3O_5$, but these two azid are not the same compound. Predict the structures of both azides a explain which is formed from each epoxide, and why.

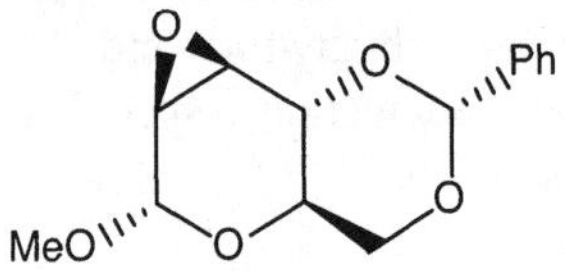

A 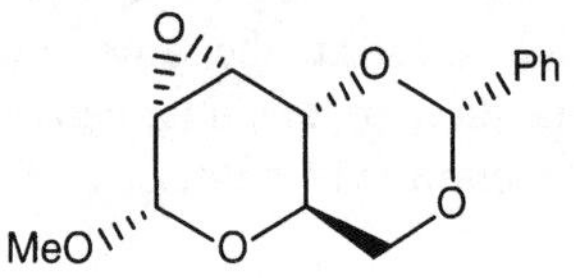   B

3. Explain the following reaction sequence. Give mechanisms for all t steps.

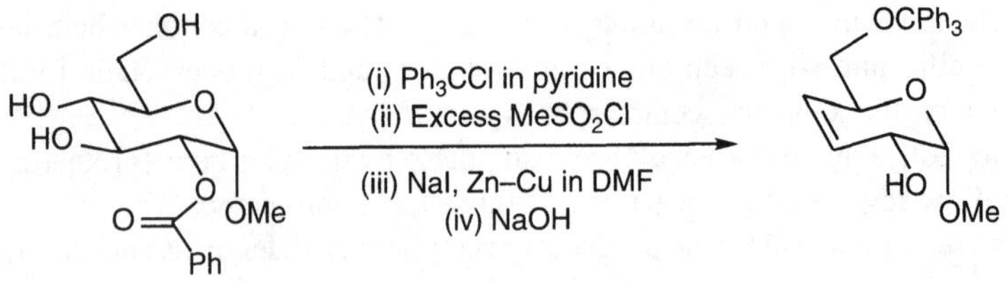

(i) Ph₃CCl in pyridine
(ii) Excess MeSO₂Cl

(iii) NaI, Zn–Cu in DMF
(iv) NaOH

5 Reactions of the anomeric centre Part II

5.1 Mutarotation and furanose/pyranose equilibration

This chapter will focus on reactions that are particular to the anomeric centre. The anomeric centre of a 'free sugar' is a hemiacetal. We saw in Chapter 3 how this cyclic hemiacetal can open, with acid or base catalysis, to allow both *mutarotation* and inter-conversion between *furanose* and *pyranose* forms, but since these processes are fundamental to any consideration of the reactivity of the anomeric centre it is worth briefly recapping. For example, the α form of glucose in the pyranose form (α-D-glucopyranose) **5.1** may ring open via use of the anomeric oxygen lone pair to produce the open chain hydroxy aldehyde **5.3**. One of several ring-closing processes can then occur. The first possibility is that the newly liberated 5-hydroxyl group can simply re-close onto the C-1 aldehyde, regenerating the 6-membered pyranose ring. Since this hydroxyl group may attack *either face* of the aldehdye, both the α-pyranose **5.1** and β-pyranose **5.2** products may be formed. Thus the α and β forms may equilibrate via this process of mutarotation. However, the open chain hydroxy aldehyde **5.3** may also ring close by nucleophilic attack of the free 4-hydroxyl group onto the aldehyde, thus forming a 5-membered ring. Again this hydroxyl group may also attack either face of the aldehyde, leading to both the α **5.4** and β **5.5** furanose products.

The term 'free sugar' is often employed to mean a carbohydrate derivative that has a hydroxyl group at the anomeric centre, that is, which is not a glycoside

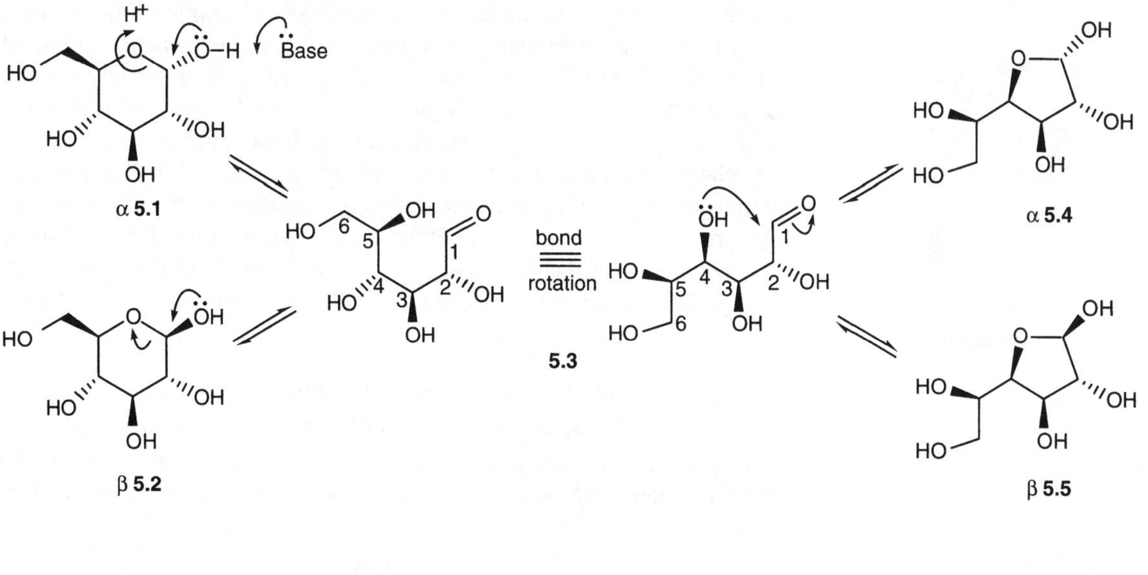

The previous chapter demonstrated how the anomeric hydroxyl group can act as a nucleophile in simple reactions such as acetylation. In this chapter we will consider the other modes of reactivity of the anomeric centre.

5.2   Nucleophilic substitution at the anomeric centre

As mentioned previously, complete acetylation is quite often the first step of synthetic sequences involving sugars. Conversion of the hydroxyl group at the anomeric position into an acetate is particularly useful since acetate can readily act as a leaving group under the appropriate reaction conditions, and, therefore, many other anomeric substituents may be introduced by nucleophilic substitution reactions. As we shall see in the next chapter, one of the primary aims of the synthetic carbohydrate chemist is the linking of two sugar units to form a disaccharide through nucleophilic displacement of a leaving group at the anomeric centre by a hydroxyl group of another sugar. It is, therefore, of paramount importance that a leaving group of one type or another can be introduced at the anomeric centre.

5.2.1   Mechanism of nucleophilic substitution at the anomeric centre—S_N1 or S_N2?

Before we go on to consider some real substitution reactions, we should consider the possible mechanisms by which nucleophilic substitution may occur. One important consideration is the stereochemistry at the anomeric centre. In fact control of the stereochemical outcome of nucleophilic substitution at this position is one of the most difficult tasks faced by the synthetic chemist.

The course of a particular nucleophilic substitution reaction is necessarily complicated by the presence of the ring oxygen, since this facilitates S_N1 type pathways. Since the oxygen has lone pairs, these can aid the departure of the leaving group and stabilise the carbonium ion intermediate by resonance (Fig. 5.1). Formally this interaction can be described as an $n \rightarrow \sigma^*$ donation of the oxygen lone pair, in a directly analogous manner to that responsible for the anomeric effect. After the loss of the anomeric leaving group L as the rate determining first step of the S_N1 pathway, the carbonium ion **5.6** is attacked by the nucleophile. However, this incoming nucleophile, Nu, can attack either face of this carbonium ion, allowing the possible formation of both the α (**5.7**) and β (**5.8**) products. As we shall see later, there are many factors, such as the nature of the solvent, or of the protecting group R on the neighbouring C-2 hydroxyl, that can play an important role in determining the relative amounts of these two products that are formed. As an extra complication it is also clear that, in certain cases, almost exclusively, S_N2 type processes can occur competitively with the concomitent clean inversion of configuration at the anomeric centre. In the hypothetical example below clean S_N2 displacement of the α-anomeric leaving group would lead exclusively to the β product **5.8**.

5.2.2   Neighbouring group participation

There are several strategies for controlling the stereochemical outcome of nucleophilic substitution reactions at the anomeric centre. As we shall see in Chapter 7, being able to selectively construct newly formed anomeric bonds,

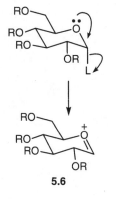

5.6

Fig. 5.1 Anomeric carbonium ions are often termed glycosyl cations

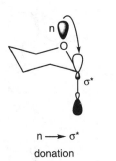

$n \longrightarrow \sigma^*$

donation

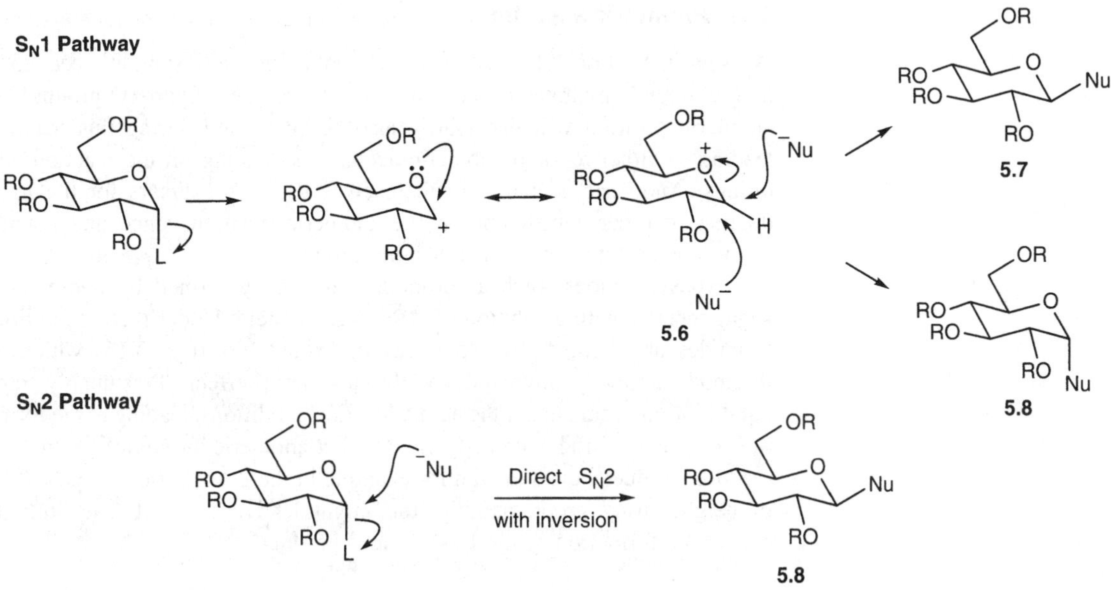

S$_N$1 Pathway

S$_N$2 Pathway

Direct S$_N$2
with inversion

either as the α- or β-anomers, is crucially important for the synthesis of di- and oligosaccharides.

One of the most useful ways to control the stereochemistry of the newly formed anomeric bond is by neighbouring group participation of an ester protecting group, such as an acetate or benzoate, on the 2-hydroxyl group. Let us consider an S$_N$1 type reaction of an acetylated sugar, which has a generalised anomeric leaving group L, with an added nucleophile Nu. After the first step consisting of loss of the anomeric leaving group, we can see that the participation of the carbonyl oxygen of the acetate protecting group at the 2 position may stabilise the intermediate glycosyl cation (**5.9** or **5.10**) by cyclisation. This so formed cyclic oxonium ion (**5.11** or **5.12**) can then be opened in an S$_N$2 fashion by the external added nucleophile, with the corresponding inversion of configuration. The net result is that the newly formed anomeric linkage is necessarily *trans* to the 2-hydroxyl group. Therefore from a *gluco* material we form specifically the β-glucoside **5.13**, whereas using a *manno* starting material, which has the epimeric hydroxyl group at the 2 position, we form, specifically, the α-isomer (mannoside) **5.14**.

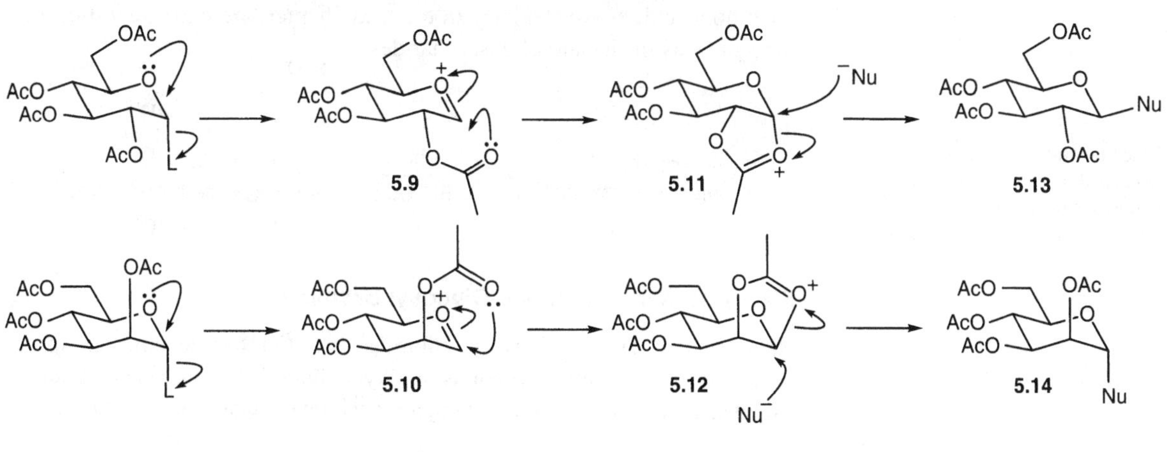

5.3   Anomeric acetates

We saw in Section 4.1 that the starting point for many synthetic sequences involving carbohydrates is acetylation of all the sugar hydroxyl groups. The anomeric position will necessarily also be acetylated during this reaction, leading to either α- or β-anomeric acetates depending on the reaction conditions. Anomeric acetates are themselves useful precursors for the introduction of other substituents at the anomeric position, since under acidic conditions acetate can act as a leaving group.

Glycosyl halides, such as bromides, are easily formed by treatment of sugar acetates with a solution of HBr in acetic acid. Under these conditions, bromides are formed exclusively as the α-anomers (e.g. **5.15**) which are thermodynamically favoured by the anomeric effect. Presumably rapid equilibration occurs under the acidic reaction conditions, leading to the formation of the thermodynamically favoured α-anomeric bromides as the sole reaction products, even though for example in the *gluco* series the possibility of neighbouring group participation might lead one to, at least initially, expect the formation of the less stable β product.

Glycosyl chlorides and fluorides can be formed analogously although the increasing strength of the halogen–carbon bond means that equilibration becomes a slower process allowing the observation of the kinetic β products in these instances.

Displacement of glycosyl acetates with other nucleophiles occurs under Lewis acidic conditions. For example, treatment of anomeric acetates with thiols, (e.g. thiophenol) in the presence of boron trifluoride etherate $(BF_3 \cdot Et_2O)$ is a good synthetic route to thioglycosides. The selenium versions—selenoglycosides—may be formed in a similar manner. Both these types of materials are extremely useful as stable anomeric substituents which can be selectively activated into good leaving groups under certain reaction conditions and, as we shall see in the next chapter, are extremely useful for the synthesis of di- and oligosaccharides.

Note that neighbouring group participation ensures that the thioglycoside is formed as the 1,2-*trans* isomer—on this occasion the β-gluco isomer.

5.4   Anomeric Halides—Glycosyl Bromides

Halides such as bromides or chlorides are familiar leaving groups in nucleophilic substitution reactions of alkyl halides. They are also particularly useful as leaving groups at the anomeric position of sugars, and their reactivity follows the usual trend in that the iodides are the most reactive, and the fluorides are the least reactive. In fact anomeric iodides are rather too reactive, and due to their high instability have not been used to a great extent in synthesis. We shall consider the use of the rather stable glycosyl fluorides, specifically in the context of the formation of disaccharides in the next chapter. Bromides are the most commonly used anomeric halides, and are easily formed, as already outlined, by treatment of the anomeric sugar acetates with a solution of HBr in acetic acid. They undergo an extremely useful and diverse array of reactions, as outlined in Fig. 5.2.

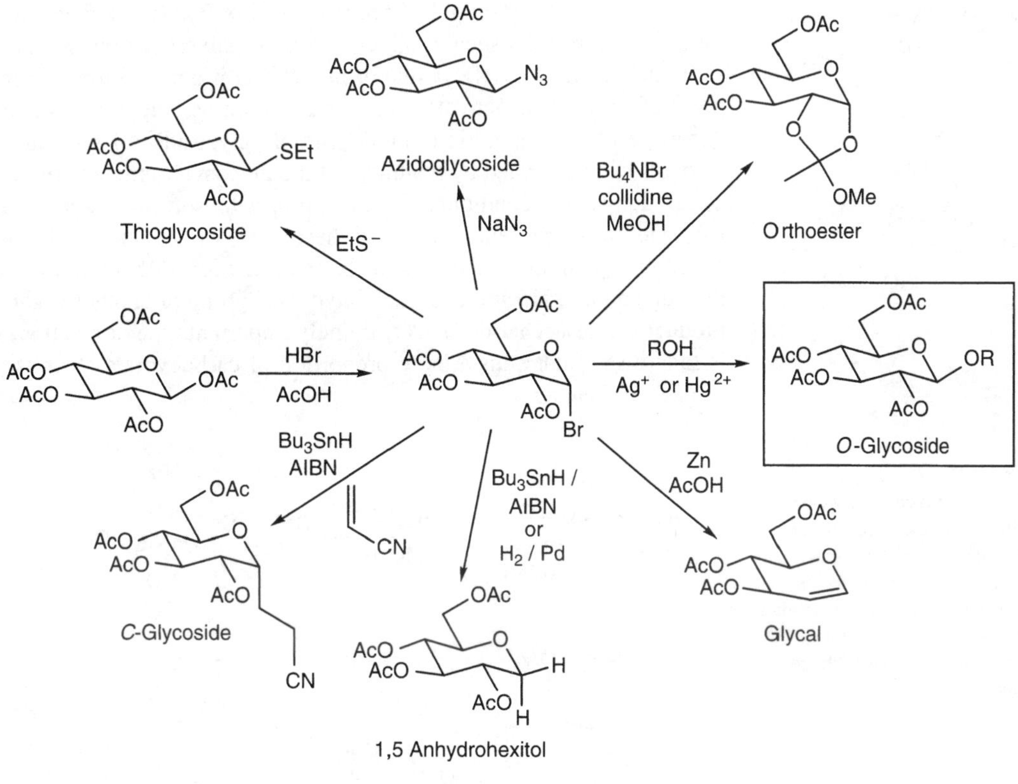

Fig. 5.2 Synthesis and representative reactions of glycosyl bromides.

5.4.1 Nucleophilic substitution reactions

Glycosyl bromides react readily with good nucleophiles. For example, reaction with thiolate anions allows the formation of thioglycosides, and represents an alternative synthesis to that from the corresponding glycosyl acetate.

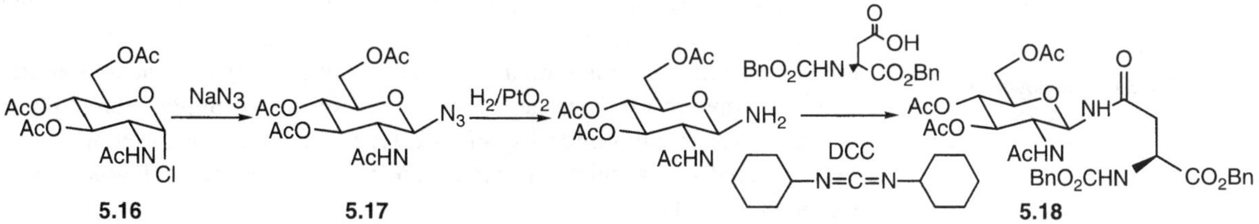

Introduction of nitrogen at the anomeric centre can be readily achieved by displacement of glycosyl halides, with either primary amines, or preferentially by nucleophilic displacement with azide, to yield azidoglycosides. For example, reaction of the glycosyl chloride of the *N*-acetyl glucosamine derivative **5.16** yields the glycosyl azide **5.17**. Subsequent reduction of the anomeric azide to produce the corresponding glycosyl amine by hydrogenation, can be followed by coupling to the β-carboxylate of a suitably protected aspartic acid derivative via standard peptide coupling techniques, and this allows access to *N*-linked glycopeptides such as **5.18**.

Reaction of glycosyl bromides with poorer nucleophiles such as alcohols necessitates the addition of halophilic activators. The best known variants are insoluble silver salts such as silver(I)oxide or silver(I)carbonate, or mercury salts such as mercury(II)cyanide or mercury(II)bromide. From a mechanistic point of view these activators can be regarded as halophiles which aid the departure of the anomeric bromide. Simple glycosides of non-carbohydrate alcohols can be formed by treatment of the glycosyl bromide with the alcohol under anhydrous conditions in an appropriate solvent, such as dichloromethane, in the presence of one of these types of activators. Obviously this conceptually simple reaction is the basis of disaccharide formation, for when the nucleophilic alcohol is in fact a hydroxyl group of another sugar then the product is a disaccharide. This extremely important type of reaction, which it is fair to say underpins a large proportion of carbohydrate chemistry, is the subject of Chapter 7.

The use of Ag$_2$O or Ag$_2$CO$_3$ as activators for the reaction between glycosyl halides and alcohols dates in fact from as long ago as 1901. The reaction is commonly known as the Koenigs–Knorr synthesis; named after the two pioneering carbohydrate chemists who first performed these investigations.

collidine
(2,4,6-trimethylpyridine)

If the glycosyl bromide possesses ester protection of the 2-hydroxyl group, we expect these types of reactions to proceed with neighbouring group participation leading to 1,2-*trans* products. However, if the glycosyl bromide is reacted with a simple alcohol, such as methanol, in the presence of a hindered base, such as collidine, then a different type of reaction can occur. Under these basic conditions the added alcohol can trap out the intermediate cyclic oxonium ion, **5.11**, leading to the formation of an orthoester **5.19**. A small amount of tetrabutylammonium bromide is added to these reactions to act as a nucleophilic catalyst and allow conversion of the initial α-bromide to the more labile β-bromide by an S$_N$2 type reaction. This β-bromide can cyclise rapidly via neighbouring group participation, to yield the cyclic oxonium ion **5.11**.

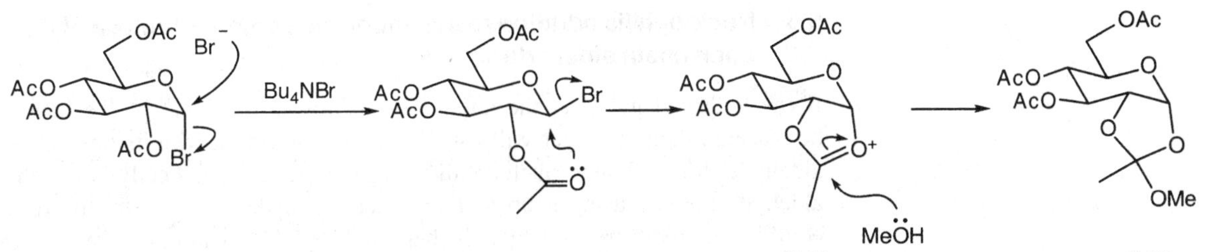

5.11   **5.19**

5.4.2 Reductive elimination reactions: synthesis of glycals

Dissolving metal reduction of peracetylated glycosyl bromides, typically with zinc in acetic acid, occurs with simultaneous elimination of the 2-acetate group to produce a cyclic carbohydrate enol ether termed a glycal, for example, galactal, **5.20**. These reactions are generally high-yielding and glycals themselves are extremely useful synthetic intermediates in particular for the synthesis of di- and oligosaccharides, as discussed in Chapter 7.

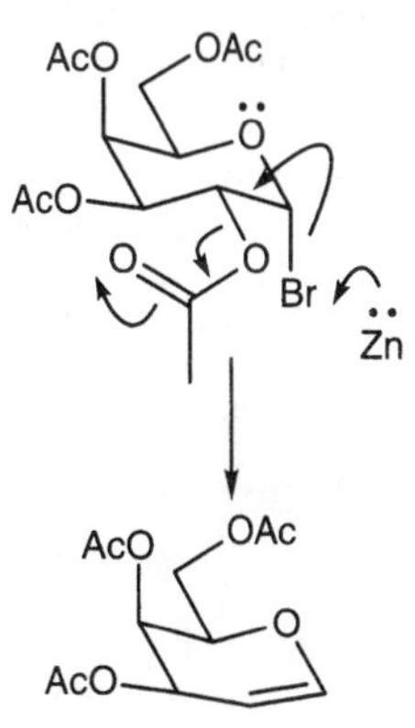

5.4.3 Free radical reactions

Finally, we should also mention free radical reactions of anomeric bromides. Use of free radical chain pathways can be a good way of making C–C bonds at the anomeric centre, thereby allowing access to *C*-glycosides. Thus, reaction of an anomeric bromide with tributyltin hydride and AIBN in the presence of an electron deficient alkene such as acrylonitrile allows the selective formation of α-*C*-glycosides (path **b**), such as **5.21**, in good yield, via a free radical chain process. In the absence of an electron deficient alkene as a coupling partner, simple reduction occurs (path **a**), by reaction of the glycosyl radical directly with the Bu_3SnH to yield the anhydroalditol **5.22**.

5.20

A *C*-glycoside is a sugar which possesses a carbon substituent at the anomeric centre

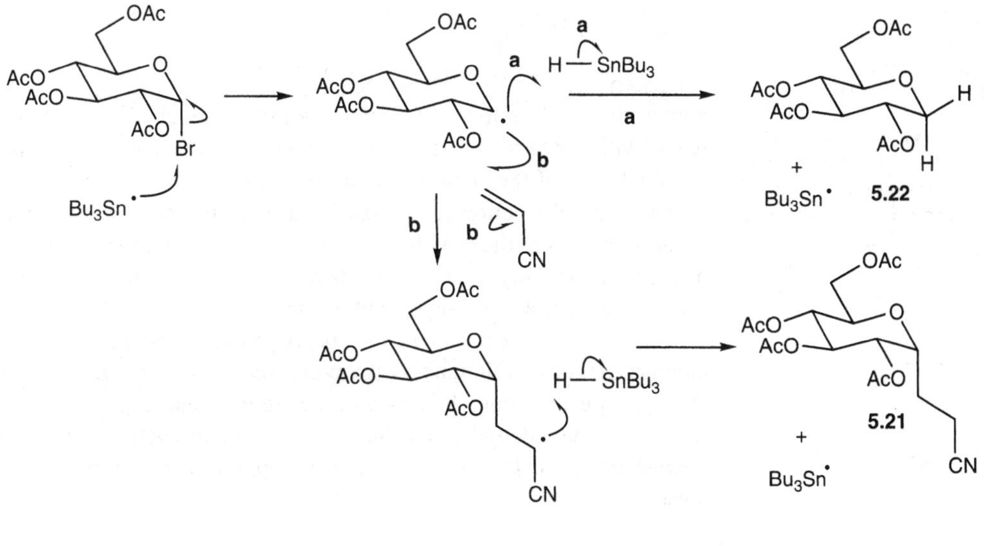

5.5   Nucleophilic addition to the anomeric centre—capture of the open chain aldehyde

If the hydroxyl group at the anomeric position is free, that is, we have a hemiacetal, then the sugar will usually be in equilibrium with its open chain aldehyde form. Many nucleophilic reagents react specifically with this aldehyde thereby using it up and so, since aldehyde and hemiacetal are in equilibrium, more free aldehyde is then produced. This can in turn react with yet more added nucleophile and so on. The overall result is that the sugar reacts as if it were completely in the aldehyde form, even though, in fact, there is only a very small amount of this open chain material present at any one time.

Such reactions are particularly important for the formation of C–C bonds at the anomeric position and, therefore, for lengthening of the carbon chain. We will consider several of the most important from a synthetic point of view.

5.5.1   Dithioacetal formation

Treatment of sugars with a free anomeric hydroxyl group with thiols, for example, ethanethiol (EtSH), in the presence of acidic catalysis, results in the irreversible formation of the open chain dithioacetal such as **5.23**. As outlined

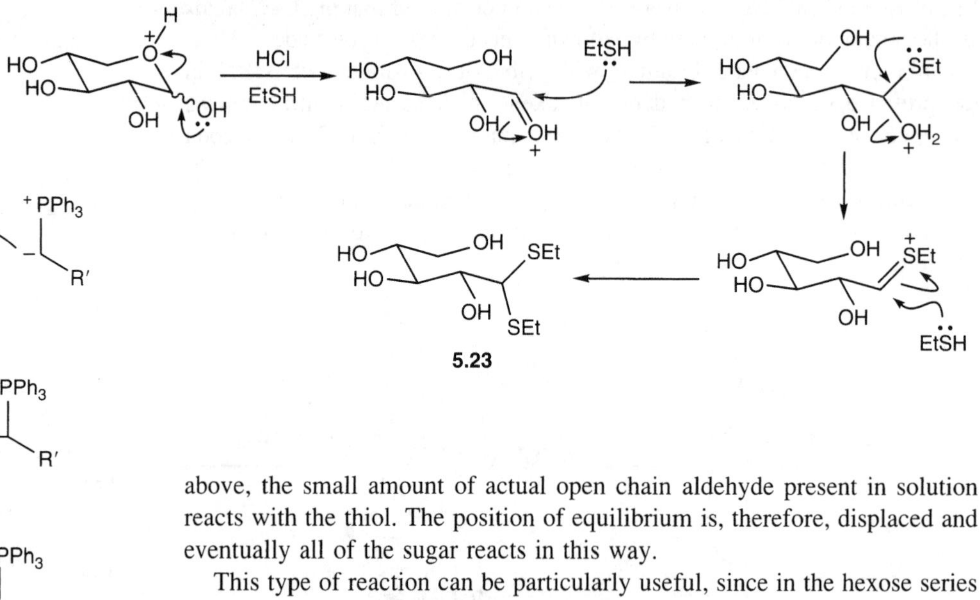

5.23

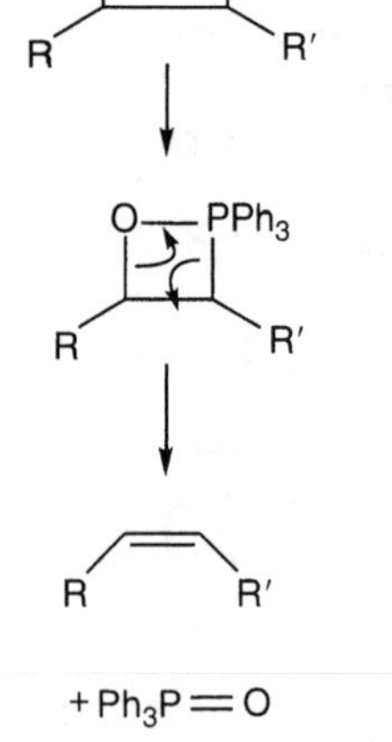

above, the small amount of actual open chain aldehyde present in solution reacts with the thiol. The position of equilibrium is, therefore, displaced and eventually all of the sugar reacts in this way.

This type of reaction can be particularly useful, since in the hexose series it allows access to the 4-hydroxyl group that is normally tied up in formation of the pyranose ring. Likewise for pentoses which exist in the furanose form it allows access to the 3-hydroxyl group.

The whole process can in fact be reversed in the presence of aqueous mercury(II) ions, which act as thiophiles, effectively pulling the reaction in the opposite direction. In this way the dithioacetal may be hydrolysed to regenerate the aldehyde, and this will once again cyclise to the appropriate hemiacetal, provided of course, that an appropriate hydroxyl group is still present!

Fig. 5.3

5.5.2 Wittig/Horner–Emmons reactions

Another class of reactions that fall into this category are the Wittig types. Once again the basics of the reaction involve attack of the added reagent, on this occasion a phosphorus ylid, onto the small amount of free aldehyde in solution. Reaction proceeds via the usual Wittig-type mechanism, via a betaine, to yield an alkene product (Fig. 5.3). Depending on the particular ylid used, the carbon chain of the carbohydrate is extended during this process.

If the reaction is performed with a stabilised ylid derived from an ester then the initial reaction product such as **5.24** is itself an α/β unsaturated ester. This material can then undergo further reaction, since the free hydroxyl group can perform a Michael addition to the double bond to yield, after proton transfer, the cyclic ether **5.25**, usually as an anomeric mixture of products. Such reactions are important routes to *C*-glycosides.

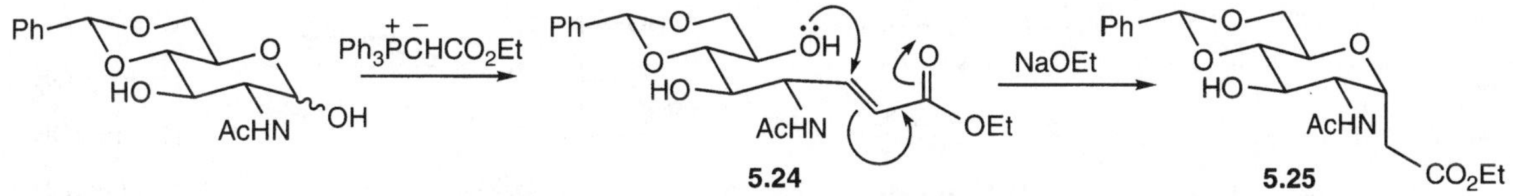

5.6 Chain degradation of sugars

As well as forcing the sugar into an open chain form and allowing access to particular hydroxyl groups which are usually involved in acetal formation, access to dithioacetals (such as **5.26**) also permits several other important reactions. One of the most notable of these involves subsequent oxidation of the dithioacetal, for example, **5.26** to a disulfone **5.27**. This has the effect of converting the terminal carbon atom (originally the aldehyde/anomeric carbon) into a leaving group, since the negative charge that would be placed on this carbon atom after its departure is stabilised by two sulfone groups. Therefore, treatment with a base can induce a fragmentation reaction to produce an aldehyde product **5.28**, which contains one less carbon atom than the starting sugar, together with the eliminated disulfone anion. Such a reaction sequence, therefore, constitutes a useful method of shortening the sugar chain by one unit and can be a very effective way of descending for example from a hexose to a pentose, in the case shown from D-glucose to D-arabinose (**5.28**).

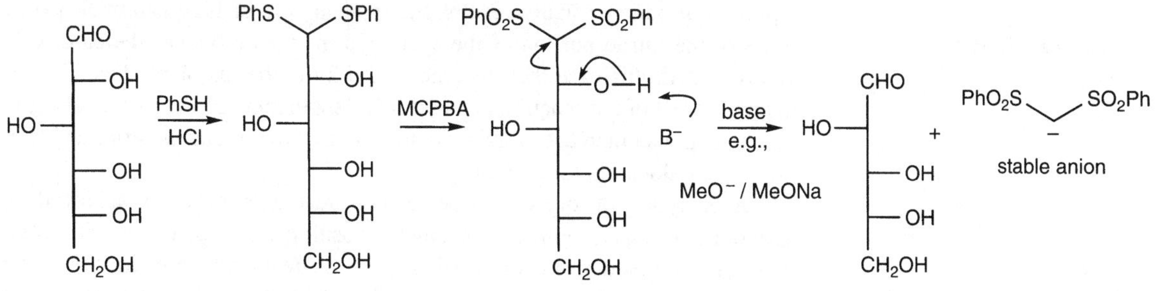

5.7 Chain extension of sugars—the Kiliani Ascension

Chain extension of a sugar by one carbon unit, that is, the direct opposite of the above degradation reaction, is a particularly important transformation for accessing so-called 'higher-carbon sugars' possessing seven or more carbon atoms. The classical **Kiliani ascension** of sugars involves treatment of a free sugar (i.e. one with a free hydroxyl at the anomeric position such as D-arabinose (**5.28**) with cyanide. Since the sugar has a free hydroxyl at the anomeric centre any ring forms are in equilibrium with the open chain aldehyde. Nucleophilic attack on the open chain aldehyde by cyanide first

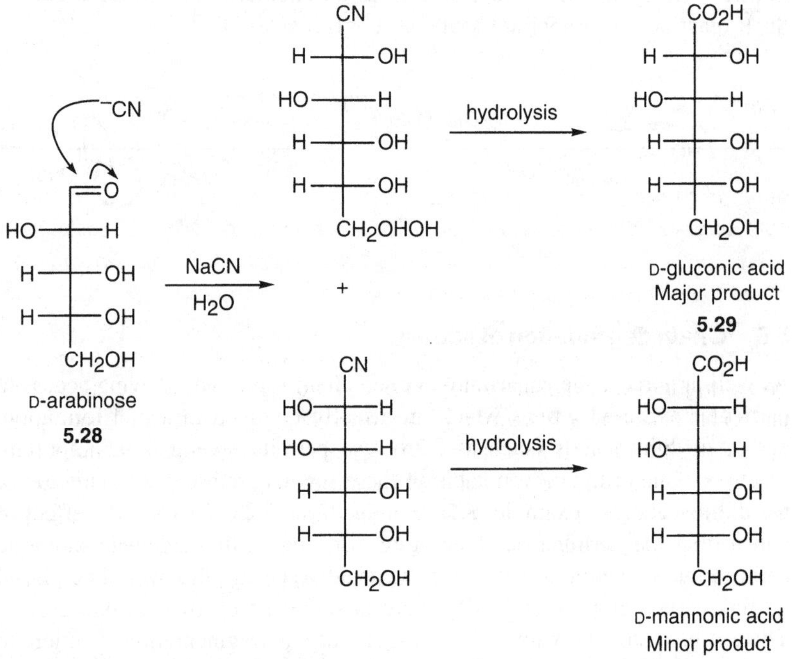

produces a cyanohydrin. The reaction proceeds to completion since the position of equilibrium is displaced and eventually all of the sugar reacts in this way. However, since the two faces of the aldehyde are diastereotopic, and either can be attacked by the nucleophile, then an unequal mixture of epimeric products usually results. Subsequent acid or base catalysed hydrolysis of the nitrile portion of the cyanohydrins then produces the carboxylic acids, and in this case the gluconic acid **5.29** is the major product. The net overall result of the reaction is therefore a lengthening of the carbon chain by one unit and an increase in the oxidation state of the 'reducing terminus' from that of an aldehyde to an acid.

Onic acids have an acid at C-1.

Other types of one-carbon extension reactions may be achieved by reactions of sugars with free anomeric positions. For example, with either the anion derived from nitromethane, or by treatment with diazomethane. Again the sugar effectively reacts as though it were all in the open chain aldehyde form.

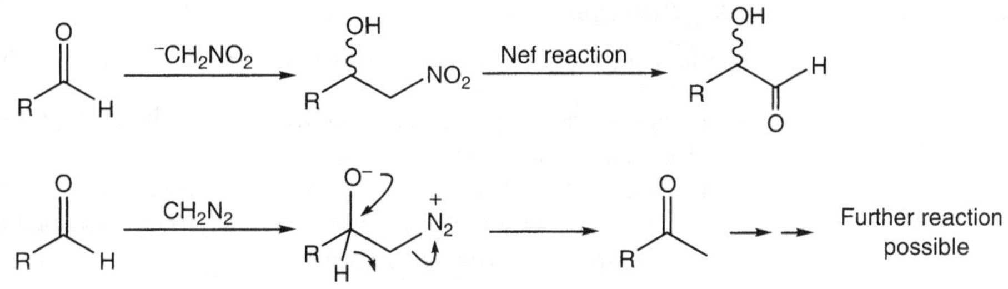

5.8 Oxidation/reduction reactions

5.8.1 Oxidation

The formal oxidation state of the anomeric centre is that of an aldehyde, and we therefore expect to be readily able to oxidise this up to the carboxylic acid level. Selective oxidation of the anomeric centre may be achieved by reaction with halogens in water. The most commonly used is bromine and the product can either be isolated as the acid salt, or, if the reaction is performed under mildly acidic conditions, then, as the corresponding lactone. The use of stronger oxidising agents such as permanganate or dichromate usually causes oxidative degradation due to the presence of other oxidisable hydroxyl groups.

The two classic tests for sugars with free anomeric centres, also termed **reducing** sugars, both involve oxidation processes. The **silver mirror test** involves treatment with Tollen's reagent, which contains Ag^+. After oxidation of the sugar to the acid, a deposit of metallic silver is observed. Similarly treatment of a sugar with Fehling's solution, which contains blue Cu^{2+}, results in the oxidation of the sugar and formation of an insoluble precipitate of red Cu(I) oxide.

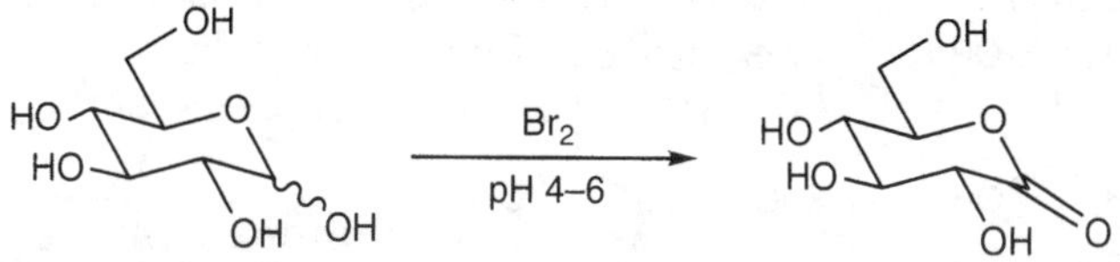

5.8.2 Reduction

Treatment of sugars with good reducing agents, such as sodium borohydride (or lithium aluminium hydride for partially protected derivatives) produces the open chain alditol in good yield. These reduction reactions proceed as expected via reduction of the small amount of open chain aldehyde in solution with, once again, the corresponding shift in equilibrium.

Note that we have already seen how anomeric halides, such as bromides may be reduced by free radical methods to leave the ring intact, and produce cyclic anhydroalditols such as **5.22**.

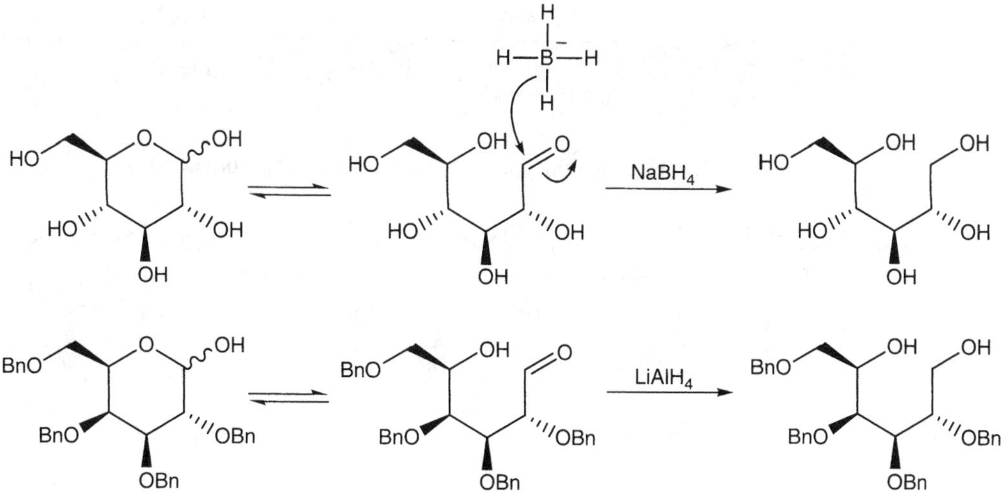

5.9   Summary

At the end of this chapter you should be able:

- to understand the possible mechanisms by which nucleophilic substitution at the anomeric centre may occur;
- to describe the problem of control of stereochemistry at the anomeric centre, and how the use of neighbouring group participation allows the formation of 1,2-*trans* glycosides;
- to know how to form anomeric bromides, and be familiar with their substitution reaction chemistry;
- to understand nucleophilic addition reactions to sugars, which occur via trapping of the free aldehyde and displacement of the ring opening equilibrium;
- to recall a method for extending the carbon chain of a sugar by one unit;
- to recall a method for shortening the carbon chain of a sugar by one unit;
- to understand the possibility of either oxidation or reduction of the anomeric centre, either in the cyclic or open chain forms.

5.10   Questions

1.   How would you convert glucose into the following compounds?

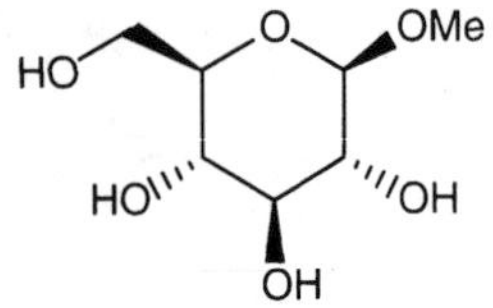

(i) β-methylglucopyranoside 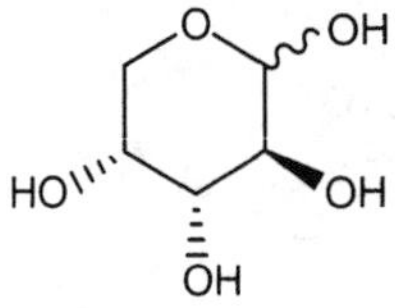   (ii) arabinose 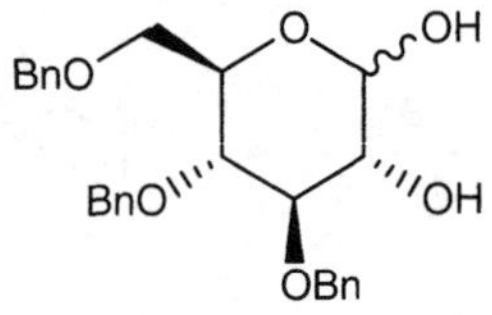   (iii) 3,4,6-*O*-benzylglucose 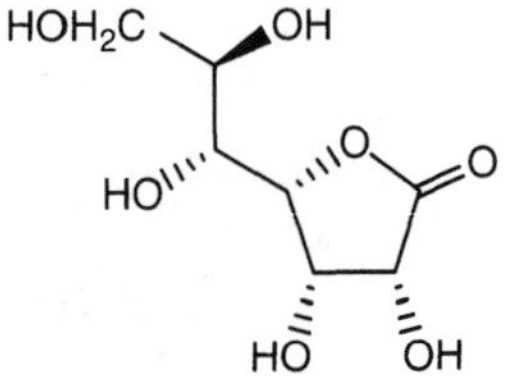   (iv) α-glucoheptonolactone

2.   Explain the following reaction sequence giving mechanisms for all steps, and predict the structure of the final product.

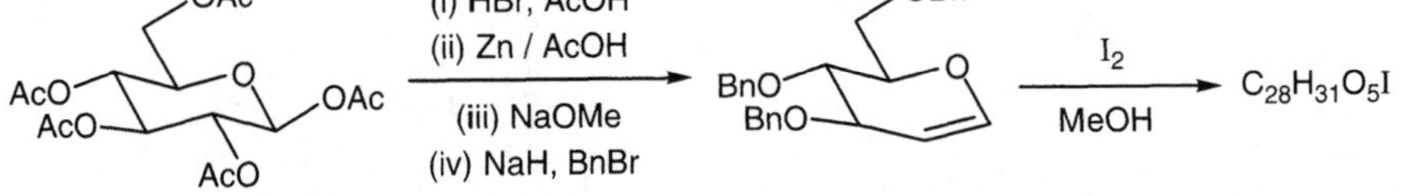

(i) HBr, AcOH
(ii) Zn / AcOH
(iii) NaOMe
(iv) NaH, BnBr

I_2 / MeOH → $C_{28}H_{31}O_5I$

3.   How would you achieve the following conversions?

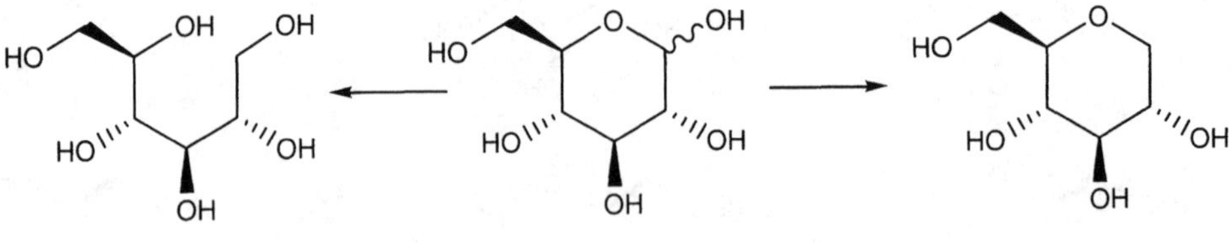

6 Reactions of hydroxyl groups Part II: cyclic acetals

We saw in Chapter 3 how reaction of carbonyl groups with alcohols under acidic catalysis can lead to the formation of acetals. This is a reversible process and the position of equilibrium depends on the reaction conditions. We have already seen how, since carbohydrates themselves possess both these functional groups, they can exist as cyclic hemiacetals, either in the 5-ring (furanose) or 6-ring (pyranose) forms. In this chapter we will focus on the reactions of carbohydrate hydroxyl groups with other aldehydes and ketones to form cyclic acetals, which most commonly find utility as a means of selectively protecting particular sugar hydroxyl groups.

6.1 Cyclic acetal formation: 5-ring versus 6-ring products

Since carbohydrates possess several hydroxyl groups, reaction of a sugar with an aldehyde or ketone under acidic conditions usually results in the formation of a cyclic acetal. The mechanism of the reaction is identical to the one we came across previously for lactol formation in Chapter 3, and occurs via an S_N1 pathway. The only difference is that since the two alcohol groups are in fact part of the same molecule, the final step is a cyclisation reaction.

cyclic acetal

S_N1 process

Strictly the product of aldehyde with diol is an acetal and of ketone with diol is a ketal. In this chapter we will tend to use acetal to describe both.

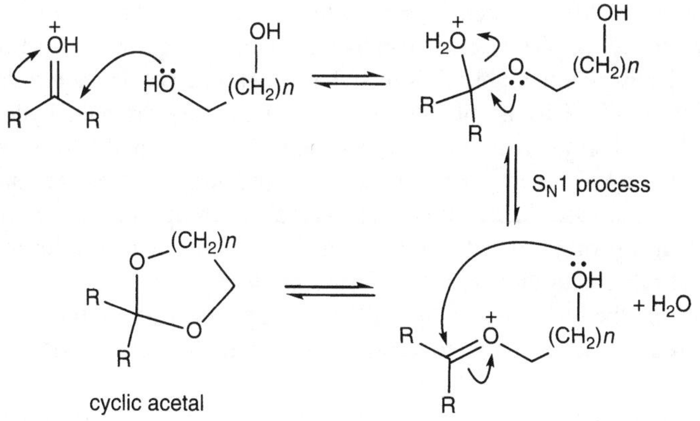

When there are several different hydroxyl groups to choose from, a mixture of different products containing different ring sizes could, in theory, be formed. For example, if we consider the reaction of glycerol **6.1** with acetone, then reaction could either lead to the 5-ring containing product **6.2**, or the alternative 6-ring containing product **6.3**. In practice almost all of the

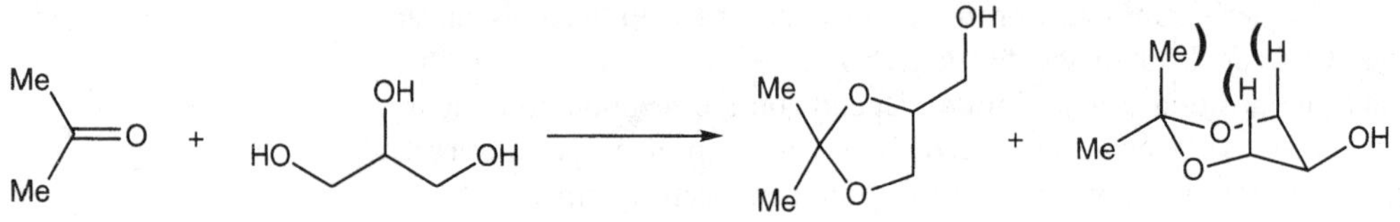

product formed is the 5-ring containing material **6.2**. We can rationalise this result by examination of the chair conformation that would be adopted by the 6-ring product **6.3**, in which one of the methyl groups would have to adopt an axial position. This would destabilise this form due to an unfavourable 1,3-diaxial repulsion between the axial methyl group and the axial hydrogen atoms. This simple type of analysis leads us to the conclusion that on thermodynamic grounds the 5-ring cyclic acetal will be the more favoured product, and indeed this is found in practice.

Remember that since this reaction is reversible then it is simply the product stability that determines product distribution. Thus any **6.3** formed would revert to **6.2** under the conditions of the reaction.

6.1 major product **6.2** minor product **6.3**

6.4 **6.1** minor product **6.5** major product **6.6**

Let us consider the reaction of glycerol **6.1** with benzaldehyde **6.4**. Again two possible products can be formed, namely the 5-ring compound **6.5** and the 6-ring product **6.6**. In contrast to the reaction of acetone, consideration of the chair conformation that would be adopted by the 6-ring product **6.6** now leads us to a different conclusion as to which would be the most stable product. In **6.6** the phenyl group can happily adopt an equatorial position, and so avoid any unfavourable 1,3-diaxial interactions. **6.6** is therefore more stable than **6.5**. This is, of course, is possible since benzaldehyde is an aldehyde, and so the axial position is occupied by the small hydrogen atom rather than by a larger group. Since in the previous example acetone is a ketone, in that case there is no choice—an alkyl group must sit in the axial position if a 6-ring product is formed, and therefore the 6-ring form is disfavoured relative to the alternative 5-ring product.

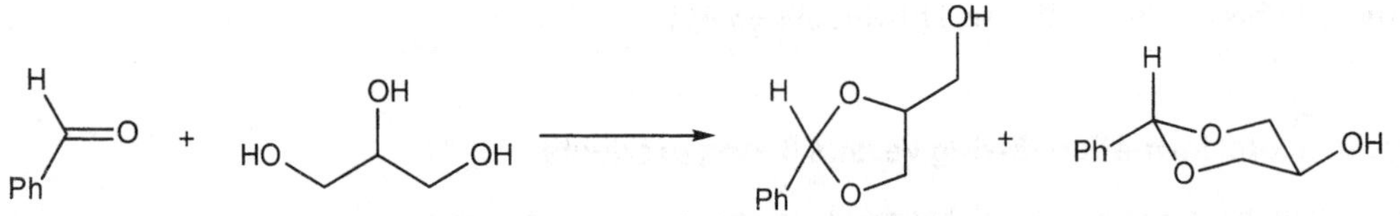

We can, therefore, come up with the following rule of thumb, which although not *always* true is nonetheless an extremely useful guide.

Ketones react with compounds possessing several hydroxyl groups to preferentially form 5-ring cyclic acetal products, whereas aldehydes react to preferentially form 6-ring cyclic acetal products.

We will now consider the utility of this differentiation when considering selective protection of pairs of carbohydrate hydroxyl groups by the formation of cyclic acetals.

6.2 Acetonide protection of carbohydrate hydroxyl groups

Cyclic acetals which result from the condensation of the two hydroxyl groups of a molecule with *acetone* are most commonly called **acetonides** though sometimes they are more correctly referred to as isopropylidene acetals. Acetonides are extremely useful protecting groups in carbohydrate chemistry. Since acetone is a ketone, 5-ring acetonides generally form preferentially instead of alternative 6-ring products. In addition, if the two hydroxyl groups used to form the acetonide are on a ring, then in order for the new 5-membered ring to form they must be *cis* to one another, since the *trans*-fused system is too strained. Where there is the possibility of reaction with multiple hydroxyl groups, this usefully allows selective access to particular hydroxyl groups of a carbohydrate by selective protection of hydroxyl pairs.

The reaction conditions involve treatment of the diol either with acetone itself, or sometimes the dimethyl acetal of acetone **6.7** (called 2,2-dimethoxypropane), under acidic catalysis. The mechanism is identical to those we have encountered previously, and the reaction is generally considered to be under thermodynamic control. Reactions of glycosides are usually quite straightforward. The simple rule of thumb is that only hydroxyl groups that are *cis* to one another prefer to react to form cyclic 5-ring acetonides. For example reaction of the methyl galactoside **6.8** with dimethoxypropane under acidic catalysis results in the selective formation of **6.9** which contains a 5-ring 3,4-acetonide, and in which the 2 and 6 hydroxyl groups remain free.

Trans-fused bicyclic 3.4.0 or 3.3.0 ring systems are much more strained than the corresponding *cis*-fused rings and, therefore, thermodynamically disfavoured.

is more strained than

6.7

+ MeOH

Reaction of diols with dimethoxypropane occurs by an analogous S_N1 mechanism, the first step of which is protonation and loss of methanol.

6.8   **6.7**   H⁺   **6.9**

Reactions of free sugars are more complicated since in addition to acetonide formation we have competing processes of mutarotation and furanose/pyranose equilibration. We shall limit our discussion to the three best known examples to illustrate the reactions that are possible, and also to demonstrate the selectivity that can be achieved.

6.2.1 Reaction of glucose with acetone and acid

In the pyranose (6-ring) form all the hydroxyl groups of glucose **6.10** are *trans* to each other and adopt equatorial positions. As such only one cyclic 5-ring *cis* acetonide could be formed, namely that involving the hydroxyl at the anomeric centre and the 2-hydroxyl group. However, under the acidic conditions used for acetonide formation the predominant 6-ring pyranose form is in equilibrium with the open chain and 5-ring furanose forms. In particular the 5-ring furanose form **6.11** could react with acetone to form two 5-ring cyclic acetonides. This is possible because the second acetonide formed by reaction of the 5- and 6-hydroxyl groups of glucose would not be part of a ring, and, therefore, there is no longer any '*cis* requirement'. This product **6.12** in which two cyclic acetonides are formed is, therefore, the thermodynamically preferred one, and as the reaction is run under conditions of thermodynamic control is, indeed, formed preferentially. Therefore, treatment of glucose with acetone and acid leads to preferential formation of the furanose diacetonide **6.12**, which is often simply called diacetone glucose, in which only the 3-hydroxyl group remains free.

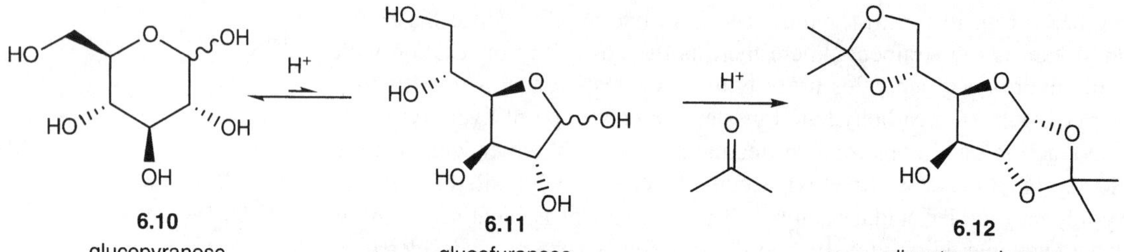

| 6.10 | 6.11 | 6.12 |
| glucopyranose | glucofuranose | diacetone glucose |

6.2.2   Reaction of galactose with acetone and acid

Galactose is epimeric to glucose at the 4-position. This means that in the pyranose form **6.13** the 3- and 4-hydroxyl groups of galactose are *cis* to each other, the 4-hydroxyl group adopting an axial orientation. Therefore reaction of galactose with acetone and acid can lead directly to a pyranose diacetonide without having to accommodate rearrangement of the preferred pyranose 6-ring into a less favourable 5-ring furanose.

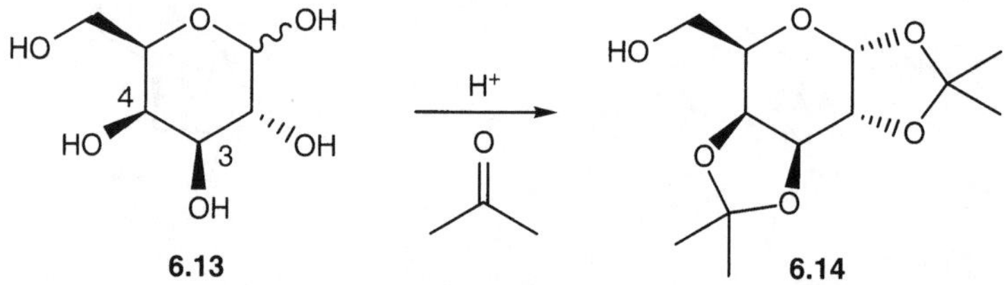

6.13	6.14
galactopyranose	diacetone galactose
3- and 4-hydroxyl groups are *cis*	

Thus treatment of galactose with acetone and acid leads to preferential formation of the pyranose diacetonide, **6.14**, which is often simply referred to as diacetone galactose. Acetonides are formed between the anomeric centre and 2-hydroxyl and also between the 3- and 4-hydroxyls, which means that only the 6-hydroxyl group of galactose remains free. It should be remembered that, again, the conditions are acidic and, again, conversion to the furanose and open chain forms is possible, but in this case the pyranose product **6.14** is the thermodynamically preferred one.

6.2.3   Reaction of mannose with acetone and acid

Mannose is epimeric to glucose at the 2-position. Therefore in the pyranose form **6.15** the 2,3-hydroxyl groups of mannose are *cis* to each other.

The result is that two possible pyranose structures, each with just one acetonide, could be formed. One would be analogous to the glucose case, a cyclic 5-ring *cis* pyranose acetonide formed between the 2-hydroxyl group and the anomeric centre, the other would be an acetonide between the *cis* 2- and 3-hydroxyls. Instead, reaction of mannose with acetone and acid proceeds similarly to glucose via the furanose form **6.16** leading to a furanose diacetonide, **6.17**, often called simply diacetone mannose. There is an important difference between the mannose and glucose cases, since in the furanose form of mannose the 2- and 3-hydroxyl groups are *cis* to each other, whereas for glucose they are *trans*. Reaction of mannose with acetone and acid can therefore result in the formation of a diacetonide in which the 2- and 3-hydroxyl groups react to form one acetonide, the other being formed by reaction of the 5- and 6-hydroxyls. This is, indeed, what happens in practice. Diacetone mannose, **6.17**, is produced in high yield simply by reaction of mannose with acetone and acid, and, unlike the previous two cases, all hydroxyl groups are protected, but the anomeric centre remains free.

6.15	**6.16**	**6.17**
mannopyranose	mannofuranose	diacetone mannose

6.3 Benzylidene protection of carbohydrate hydroxyl groups

The formation of cyclic benzylidene acetals by reaction of benzaldehyde and catalytic acid is in mechanistic and practical terms directly analogous to the formation of acetonides discussed above. Thus, reaction of a diol with benzaldehyde, or the dimethyl acetal of benzaldehyde, under acidic catalysis results in the formation of a cyclic benzylidene acetal. However, as we have already mentioned, since benzaldehyde is an aldehyde rather than a ketone we expect to preferentially form products containing 6-membered rings, which adopt a chair conformation with the bulky phenyl group in an equatorial position. In addition, because a 6-membered ring is formed, there is no longer a requirement that the two hydroxyl groups need be *cis* to each other. The net result is that the benzylidene protecting group is very selective for reaction with the 4- and 6-hydroxyls of carbohydrates, to form either *cis*- or *trans*-fused ring systems in which the phenyl group adopts an equatorial position. Thus reaction of either methyl glucopyranoside, **6.18**, or methyl galacto-pyranoside, **6.19**, with benzaldehyde dimethylacetal and catalytic acid produces an excellent yield of the 4,6-benzylidene protected materials **6.20** and **6.21**, respectively, in which only the 2- and 3-hydroxyl groups are free. Note that the phenyl group occupies an equatorial position in both the *trans*- and *cis*-fused ring systems, **6.20** and **6.21**, respectively.

Formation of a 4,6-benzylidene is also quite frequently performed by reaction of the sugar with benzaldehyde and ZnCl₂ under Lewis acid, rather than protic catalysis.

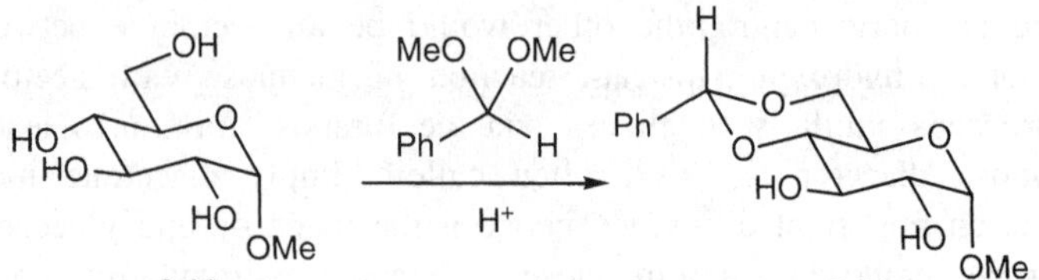

6.18

6.20

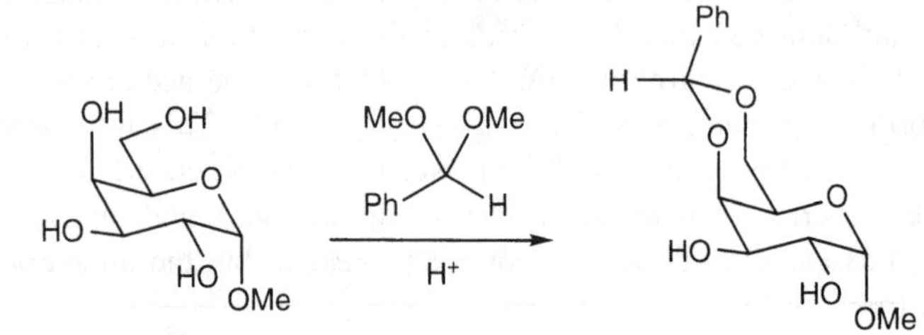

6.19

6.21

6.4   Butane diacetal protection of diols

The trimethylorthoformate provides two equivalents of methanol and activates the ketone to nucleophilic attack by the sugar hydroxyls (ROH) via an S_N1 pathway.

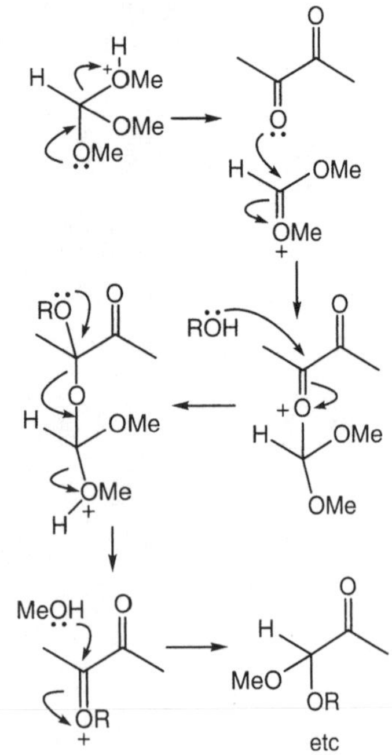

Another type of cyclic acetal protecting group has been developed more recently, which is selective for 1,2-*trans* diols, and in particular 1,2-diequatorial hydroxyl groups. This third type of cyclic acetal can, for example, be readily formed by reaction of butane-2,3-dione, **6.22**, and trimethylorthoformate with a diol under acidic catalysis, to form what is termed a butane diacetal (BDA). In many cases the reaction results in the regio- and stereoselective formation of a single cyclic acetal product. This type of protection is particularly useful for manno sugars where the equatorial 3- and 4-hydroxyl groups can be protected in the presence of both the primary 6-hydroxyl and the axial 2-hydroxyl group. For example reaction of the mannose derivative, **6.23**, with butane-2,3-dione, **6.22**, and trimethylorthoformate produces an excellent yield of the 3,4-butane diacetal protected material, **6.24**, as the sole reaction product.

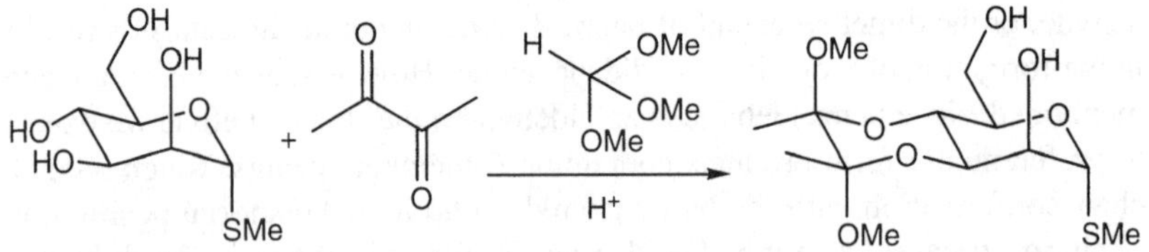

6.23   **6.22**   **6.24**

Formation of a single product is explained by the combination of two factors. First, a 6-membered diacetal ring, which adopts a chair conformation and in which the two methyl groups occupy the sterically more favoured equatorial positions, is always preferentially formed on thermodynamic grounds. In addition the two acetal methoxy groups occupy axial positions on this chair. At first this may seem rather disfavoured, but closer inspection reveals the operation of four anomeric effects (one operating in each direction at each acetal centre), providing a large thermodynamic driving force for the formation of this particular product.

6.5 Cyclic acetal hydrolysis: deprotection of carbohydrate hydroxyl groups

Since acetal formation is reversible, all the cyclic acetals that we have seen can be simply hydrolysed by treatment with aqueous acid. The presence of excess water pushes the equilibrium back to the original carbonyl compound and the corresponding diol. Although this may not at first seem a particularly useful thing to do, when coupled with other steps, such as protection of free hydroxyl groups with non-acid labile protecting groups (see Chapter 4), that are performed after initial selective acetal formation, this can be an efficient way of selectively protecting sugars. The strength of the aqueous acid used for acetal hydrolysis is an important consideration. Thus acetonides and 4,6-benzylidenes can be hydrolytically cleaved with aqueous acetic acid, or aqueous trifluoroacetic acid without disturbing the anomeric centre, which is, of course, itself an acetal and, therefore, labile to hydrolysis. Hydrolysis of the anomeric centre itself usually requires more forcing conditions, such as treatment with aqueous mineral acid.

Sometimes acetonides and benzylidenes may be cleaved selectively in the presence of each other, with, for example, aqueous acetic acid. As a general rule, primary acetonide protecting groups may be selectively hydrolysed in the presence of secondary ones. Thus, the primary acetonide of diacetone glucose, **6.12**, may be selectively hydrolysed with aqueous acetic acid to yield the monoacetonide, **6.25**, which now possesses three, free hydroxyl groups.

$2 \times n \longrightarrow \sigma^*$ donation

Both oxygen atoms have lone pairs that are antiperiplanar to the other C–O bond. Hence the anomeric effect operates in both directions, at both acetal centres, substantially stabilising this form. This $2 \times n \rightarrow \sigma^*$ stabilising effect is also seen at the other acetal centre.

As well as being hydrolysed by treatment with aqueous acid, under appropriate conditions 4,6-benzylidene acetal protecting groups may also be completely removed by catalytic hydrogenolysis like benzyl ethers. In addition they may also be selectively reductively cleaved to yield either the 4- or 6-benzyl ethers as desired (the other hydroxyl group becoming free). This flexibility greatly increases the utility of the 4,6-benzylidene acetal as a protecting group.

6.12   **6.25**

Under harsher hydrolytic conditions, such as aqueous mineral acid, all acid labile groups may be removed, including hydrolysis of any anomeric substituent. This may have consequences such as reversion to a pyranose form from a furanose form, etc. Obviously there are many, many potential reaction sequences here and we shall only be able to consider two representative examples to illustrate the sorts of overall transformations that are possible.

As an example, reaction of glucose with methanol and acid first produces the pyranose methyl glycoside, **6.18**, thereby protecting the anomeric position through a Fischer glycosylation reaction (see Chapter 3). Subsequent reaction with benzaldehyde, in the presence of zinc chloride as a Lewis acid catalyst, then produces the 4,6-benzylidene protected material, **6.20**. Reaction of **6.20**, in which now only the 2- and 3-hydroxyl groups are unprotected, with methyl iodide and base results in the formation of the dimethyl ether **6.26**. Finally, treatment of **6.26** with mineral acid results in the removal of all the acid labile protecting groups, namely the 4,6-benzylidene acetal *and* the methyl glycoside. However, the methyl ethers are stable to acid (they are ethers and not acetals!!), and therefore the final product is the dimethylated compound **6.27**, in which only the 2- and 3-hydroxyl groups remain protected.

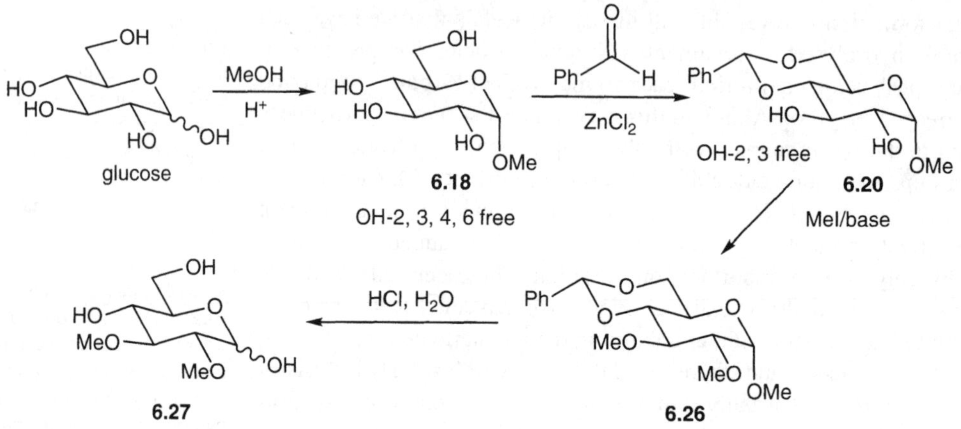

The formation of **6.29** is a second example of an overall transformation. Reaction of glucose with acetone and acid, as we have previously seen, results in the formation of the furanose diacetonide, **6.12**, in which only the 3-hydroxyl group is free. Reaction of this material with methyl iodide and base then produces the 3-*O*-methyl ether **6.28**. Finally, treatment with mineral acid results in hydrolysis of all the acetonide protecting groups, and also reversion to the more stable pyranose form. Of course, once again, the methyl ether is stable to these acidic conditions, and therefore, the structure of the final product is the pyranose methyl ether **6.29**, in which only the 3-hydroxyl group remains protected.

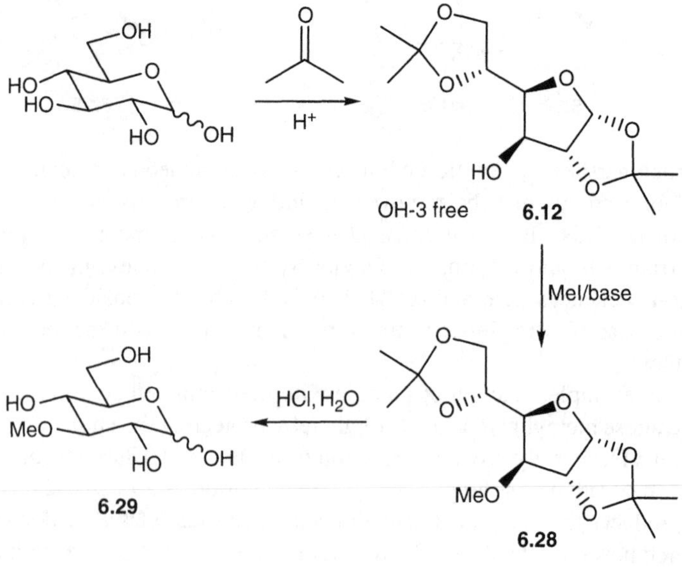

6.7 Summary

At the end of this chapter you should be able:

- to explain why the reaction of carbohydrates with acetone and acid selectively produces cyclic 5-ring acetonides, whereas reaction with benzaldehyde and acid produces 6-ring benzylidenes;
- to remember that the reaction of glucose with acetone and acid produces the furanose diacetonide with the 3-hydroxyl group free;
- to remember that the reaction of galactose with acetone and acid produces the pyranose diacetonide with only the 6-hydroxyl group free;
- to remember that the reaction of mannose with acetone and acid produces the furanose diacetonide with only the anomeric centre (i.e. the 1-hydroxyl group) free;
- to explain why protection of carbohydrates as butane diacetals results in selective protection of vicinal diequatorial diol groups;
- to be able to differentiate between acid stable and acid labile protecting groups and, hence, explain reaction sequences involving selective acetal protection, followed at a later stage by acidic hydrolysis.

6.8 Questions

1. Explain and give a mechanism for the following transformation.

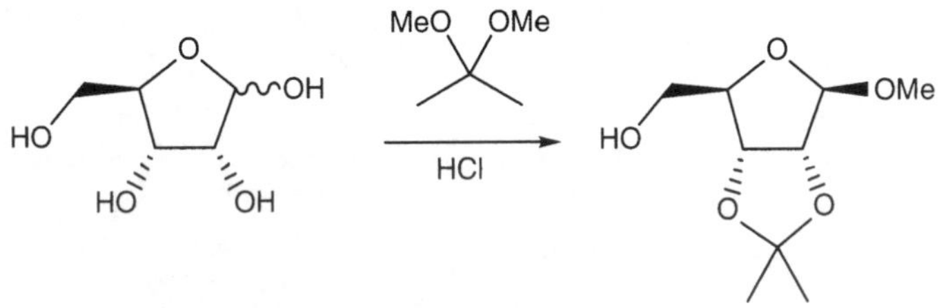

2. Predict the structures of the following products and give mechanisms for their formation.

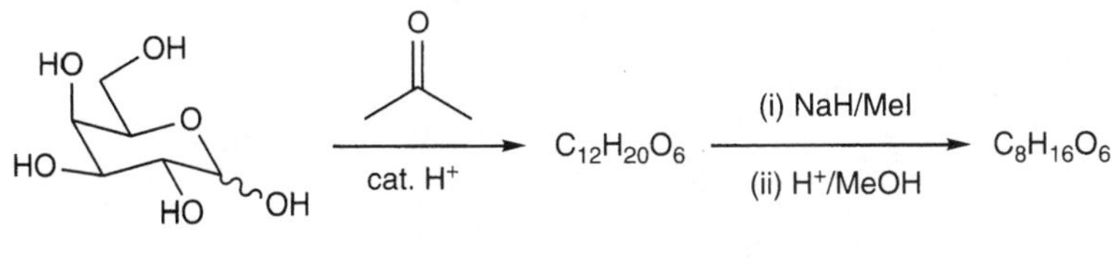

3. Suggest how you would convert glucose into the following compounds.

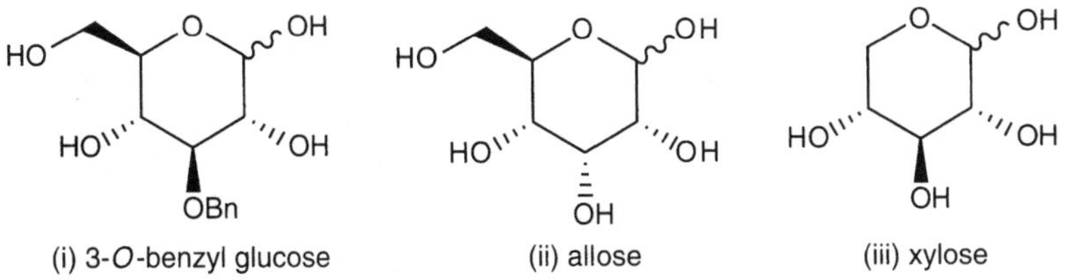

(i) 3-*O*-benzyl glucose   (ii) allose   (iii) xylose

4. Explain and give mechanisms for the following series of synthetic transformations.

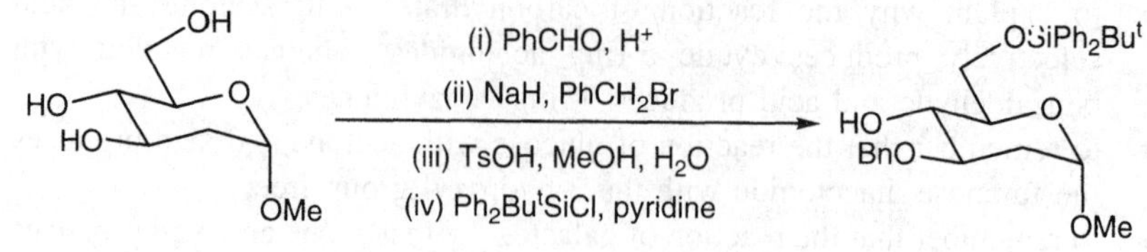

OH
HO
HO
O
OMe
(i) PhCHO, H+
(ii) NaH, PhCH2Br
(iii) TsOH, MeOH, H2O
(iv) Ph2ButSiCl, pyridine
OSiPh2But
HO
BnO
O
OMe

5. Predict the structure of the product of the following reaction. Explain why only a single reaction product is formed.

HO
OH
O
HO
HO
OMe
+
OMe
OMe
OMe
OMe
TsOH
MeOH

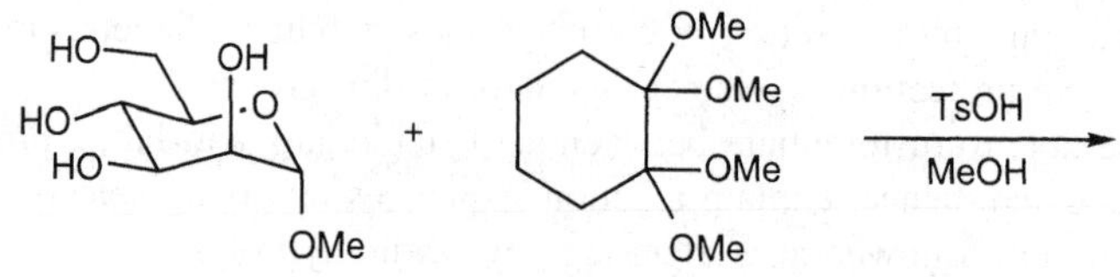

7 Chemical disaccharide formation

7.1 Glycosylation reactions

In this chapter, we shall consider the crucial step in the synthesis of any oligosaccharide, namely, the linking of two monosaccharide precursors by construction of the glycosidic linkage. Chapter 8 will consider enzyme catalysed approaches to this synthetic problem, while in this chapter we shall limit discussion to purely chemical methods.

Glycosidic bond formation is achieved by the conceptually very simple operation of displacement of a leaving group at the anomeric position of one sugar, termed the **glycosyl donor**, for example, **7.1**, with the free hydroxyl group of another, itself termed the **glycosyl acceptor**, for example, **7.2**. Such a transformation, which in this case forms a disaccharide, **7.4**, is generally termed a **glycosylation reaction**. This process may be iterated to build up larger oligosaccharides.

The **glycosyl acceptor** is the *nucleophile*.

The **glycosyl donor** is the *electrophile*.

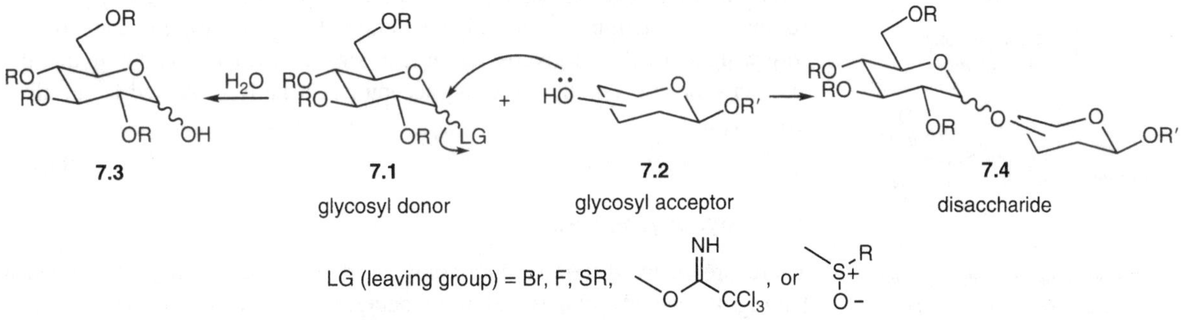

7.3	**7.1**	**7.2**	**7.4**
	glycosyl donor	glycosyl acceptor	disaccharide

LG (leaving group) = Br, F, SR,

Though conceptually extremely simple, this operation has in fact been frustrating chemists for nearly a hundred years since Köenigs and Knorr published the first chemical glycosylation reaction at the turn of the twentieth century. The problems inherent in this synthetic transformation are manifold. First, and perhaps most obviously, such reactions must be performed under completely anhydrous conditions, since the presence of even small amounts of water will lead to the formation of hydrolysis products such as **7.3**, by water reacting as a competitive nucleophile. To combat this possibility molecular sieves are nowadays frequently added to reaction mixtures in order to scavenge water. Once anhydrous conditions are found three major challenges of the glycosylation reaction remain:

Molecular sieves are zeolites (crystalline metal aluminosilicates) whose structures consist of a 3D network of silica and alumina tetrahedra. Natural water of hydration is removed from this network by heating to produce cavities which selectively absorb molecules of a specific size, most notably water (4 Å sieves). Once absorbed inside water molecules are effectively trapped, and are, thus, removed from the reaction mixture.

- regioselectivity, that is, which particular hydroxyl group of the glycosyl acceptor reacts as the nucleophile;

- stereoselectivity, that is, whether the newly formed interglycosidic linkage is specifically α or β;
- efficiency, that is, the fact that alcohols are not particularly good nucleophiles, particularly hindered secondary hydroxyl groups of partially protected glycosyl acceptors, can result in often moderate overall yields.

As we shall see the problems of regioselectivity can be overcome by selective protection of the glycosyl acceptor and we have already seen much chemistry that will enable us to do this. Although a completely general approach to total stereocontrol of the formation of any glycosidic bond remains as yet unrealised, we shall see that in many specific cases high levels of stereocontrol may be achieved by judicious choice of reaction conditions and protecting groups.

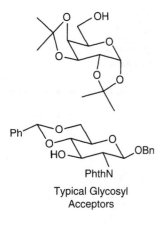

Typical Glycosyl
Acceptors

PhthN— =

Phthalimide, a useful nitrogen
protecting group, cf. the
Gabriel synthesis

7.2   Glycosyl acceptors

Glycosyl acceptors are typically monosaccharides that only have a single hydroxyl group unprotected. Although the nucleophilicity of hydroxyl groups around a sugar ring does vary slightly depending on their position and orientation, it is in general much more preferable to work with glycosyl acceptors which only have a single hydroxyl free to avoid the formation of mixtures of regioisomeric products. The other important consideration is that the anomeric position of the glycosyl acceptor should be suitably protected or differentially functionalised from the anomeric position of the glycosyl donor so that self-reaction and/or polymerisation do not occur. In many of the following examples of glycosylation reactions the reader should note that the anomeric position of the glycosyl acceptor is usually protected as an alkyl glycoside.

7.3   Glycosyl donors

There are many different, commonly used types of glycosyl donor, each having its own advantages and disadvantages. The crucial consideration here is that we must be able to activate the anomeric leaving group of the glycosyl donor and perform the nucleophilic substitution reaction, in the presence of another anomeric centre and any existing glycosidic linkages. The two essential criteria are therefore:

- the substituent at the anomeric centre of the glycosyl acceptor must remain unaffected by the activation conditions;
- any other existing glycosidic linkages, be they in the donor or acceptor, must also be unaffected by the process; otherwise oligosaccharide synthesis, wherein either the glycosyl donor or acceptor (or both) are at least disaccharides, would be impossible.

We shall now briefly survey the synthesis and activation of a few of the most commonly used types of glycosyl donors.

7.3.1 Glycosyl bromides

In Chapter 5 we came across glycosyl bromides, and indeed encountered some nucleophilic substitution reactions of these compounds. In fact, glycosyl bromides were the first glycosyl donors to be used for disaccharide formation by Köenigs and Knorr in their classic glycosylation reaction at the turn of the twentieth century.

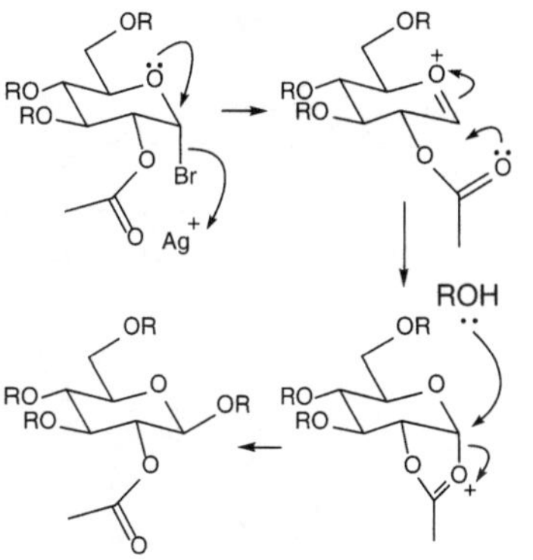

Typically, for efficient reaction with an acceptor, particularly those of low nucleophilicity, these bromides require activation, for example, by the addition of halophiles such as silver or mercury salts. The original activators used were insoluble silver oxide or carbonate (Ag$_2$O, Ag$_2$CO$_3$) but these have now been largely superseded by the use of soluble silver triflate or perchlorate (AgOTf or AgClO$_4$), or by soluble mercury salts (e.g. HgBr$_2$ or Hg(CN)$_2$). The reaction probably proceeds via an S$_N$1 type pathway. Despite the fact that they were the first examples of glycosyl donors, glycosyl bromides are still extremely widely used today, but it must be borne in mind that these reactive materials are not particularly stable—they are usually generated and then used directly for glycosylation. Note that in the example below, the newly formed disaccharide linkage is of β configuration due to neighbouring group participation of the ester protecting group on the 2-hydroxyl of the donor.

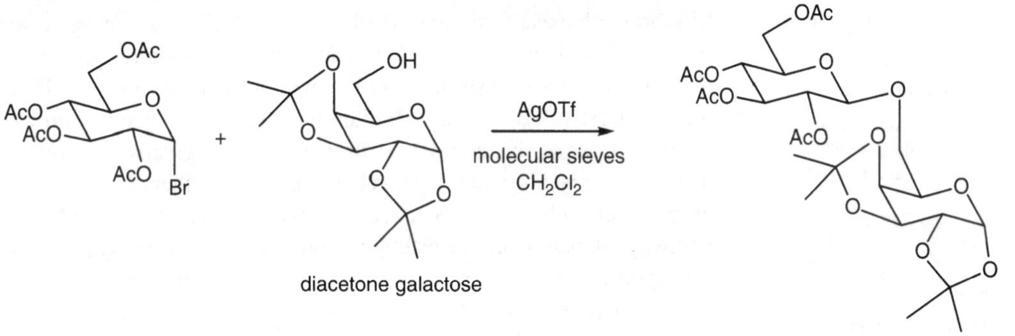

diacetone galactose

7.3.2 Thioglycosides

Thioglycosides are extremely useful as glycosyl donors since they may be readily activated under conditions which do not affect other glycosidic linkages. As we saw in Chapter 5, they are readily prepared from the corresponding anomeric acetates by treatment with a thiol and a Lewis acid, such as boron trifluoride etherate (BF$_3$ · Et$_2$O). Note again the specific formation of β-anomers due to neighbouring group participation.

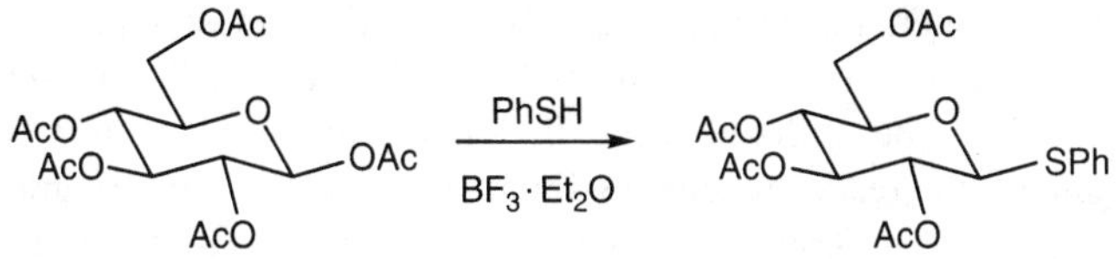

N-iodosuccinimide (NIS)

iodonium dicollidine
perchlorate (IDCP)
again basically a source of I⁺

dimethyl(thiomethyl)
sulfonium triflate
(DMTST)

a selenoglycoside

meta-chloroperoxybenzoic
acid (MCPBA)

Thioglycosides have the advantage over glycosyl bromides that they are more stable chemical species, and will not react until they are activated under particular reaction conditions. In addition, these activation conditions are particularly selective, and do not affect either existing glycosidic linkages, or the vast majority of other functional groups. Activators for thioglycosides can be basically regarded as sources of soft electrophiles. *N*-iodosuccinimide (NIS) is probably the most frequently used, often with a catalytic amount of trifluoromethanesulfonic acid (triflic acid, TfOH) and can be regarded as a source of the species I^+. Activation of the anomeric sulfur occurs by nucleophilic attack of sulfur on iodine with loss of succinimide. This converts the anomeric sulfide into a sulfonium leaving group since it is now positively charged. Loss of this anomeric sulfur produces a glycosyl cation, which can then be trapped by the glycosyl acceptor. Whether the reaction actually proceeds entirely via this S_N1 type pathway, or via a mixture of S_N1 and S_N2 pathways depends on the reaction conditions, and in many cases is unclear. Many other activators can also be used to activate thioglycosides, such as methyl triflate (MeOTf), iodonium dicollidine perchlorate (IDCP), and dimethyl(methylthio)sulfonium triflate (DMTST). Again, all are sources of soft electrophiles which convert the anomeric sulfur to a leaving group.

We should also, in passing, briefly mention selenoglycosides, which are the selenium variants of thioglycosides. These materials are very similar in many respects to thioglycosides but are more reactive when treated with an activator such as NIS. It is, therefore, possible to selectively activate selenoglycosides in the presence of thioglycosides, often a useful strategy in oligosaccharide synthesis.

7.3.3   Glycosyl sulfoxides

Glycosyl sulfoxides are amongst the most reactive glycosyl donors known and are, therefore, worthy of special mention. They are readily synthesised from the corresponding thioglycoside by oxidation, most typically with a peracid such as *meta*chloroperoxybenzoic acid (MCPBA). Two points should be noted. First, the sulfoxides are invariably produced as a mixture of diastereomers (the reader is reminded that sulfoxides are chiral) and second, it is important to avoid over-oxidation to the anomeric sulfone. Activation of anomeric sulfoxides is achieved specifically by treatment with triflic anhydride, with subsequent loss of phenyl sulfenyltriflate (PhSOTf). Addition of the glycosyl acceptor then produces disaccharides in good yield. Note in the example below the use of thioglycoside **7.5** as a glycosyl **acceptor**; thioglycosides are unaffected by these activation conditions and, therefore, constitute an **orthogonal** class of glycosyl donor.

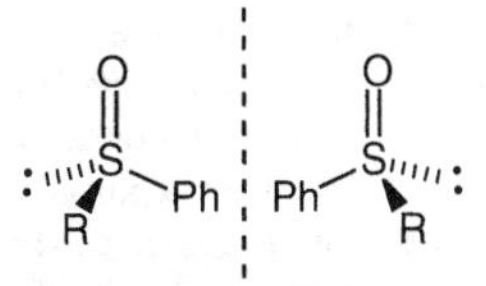

Remember disubstituted
sulfoxides are chiral

(R ≠ Ph)

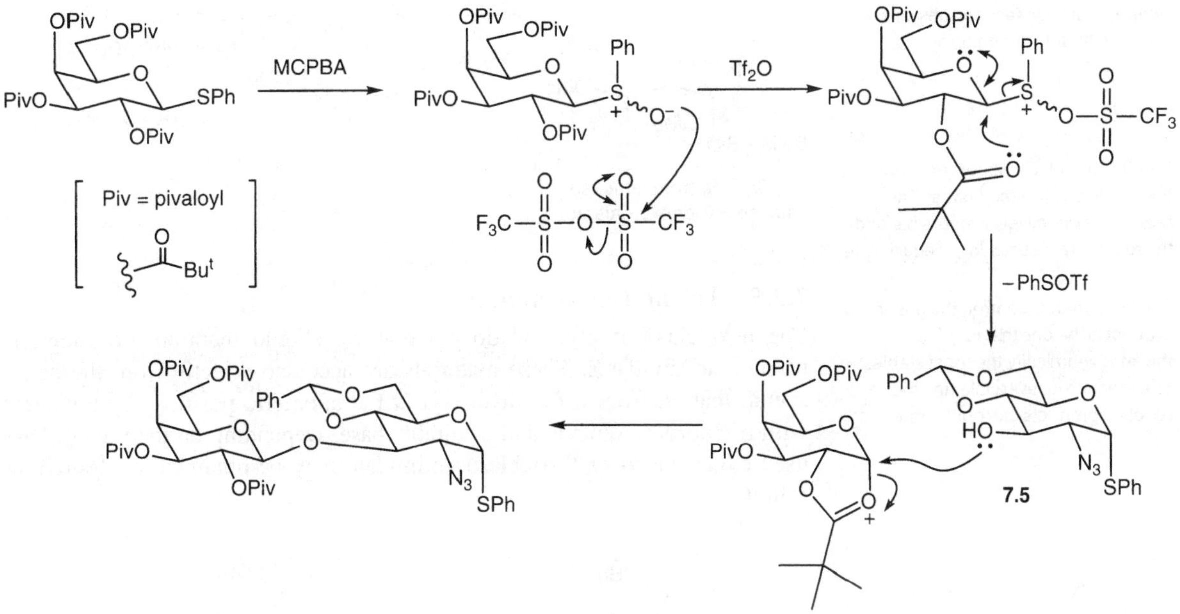

7.3.4   Glycosyl fluorides

Glycosyl fluorides are much more stable than the corresponding bromides and chlorides, and therefore offer certain advantages over the other halides. Fluoroglycosides can be readily synthesised from the corresponding thioglycosides by treatment with diethylaminosulfur trifluoride **7.6** (DAST, Et$_2$NSF$_3$). DAST is also effective for converting the free anomeric OH directly to the fluoride. Usually, α/β-anomeric mixtures of fluorides are obtained; the ratio is often solvent dependent.

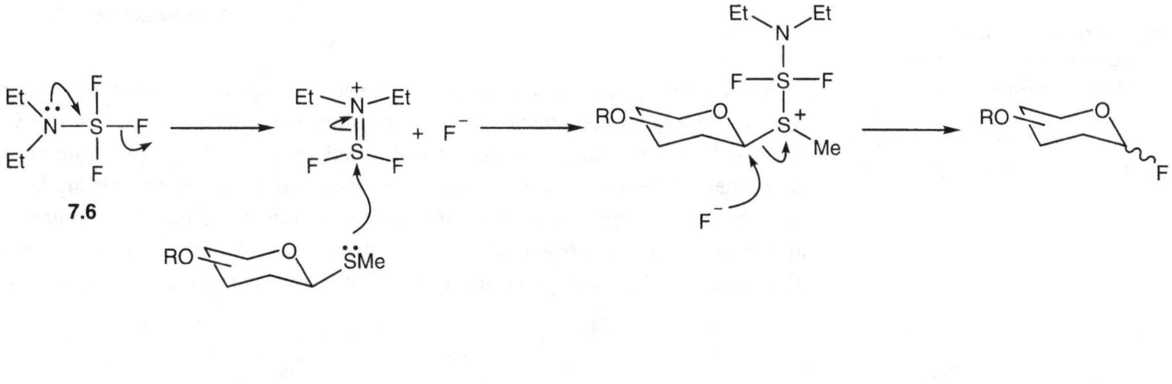

Since both fluoroglycosides and thioglycosides are unaffected by the activation conditions required for the activation of the other, a partially protected fluoroglycoside can be used as a glycosyl acceptor in a glycosylation reaction with a thioglycoside glycosyl donor, and vice versa. These are therefore termed **orthogonal** glycosyl donors. This type of combination can be extremely useful for the iterative synthesis of oligosaccharides as we shall see later in this chapter.

Activation can be performed by the addition of hard Lewis acids such as BF_3 etherate or $SnCl_2$. Particularly selective activation can be achieved with a mixture of the (bis)cyclopentadienyl dichlorides of either hafnium or zirconium (e.g. Cp_2HfCl_2) together with silver perchlorate, conditions which do not effect thioglycosides.

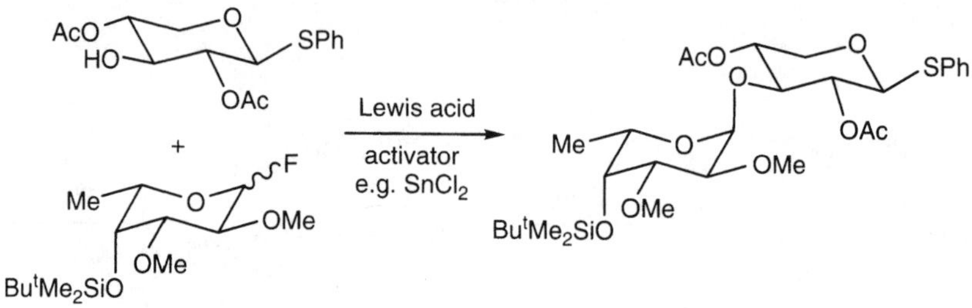

Derived from L-fucose
(fucose = 6-deoxy galactose)

Kinetic control: the observed product is the one that is formed fastest. The reaction is effectively irreversible and the reaction products do not equilibrate.

Thermodynamic control: the observed product is the one that is thermodynamically the most stable. The reaction is reversible and the reaction products can equilibrate.

7.3.5   Trichloroacetimidates

The next class of glycosyl donors that we should mention are anomeric trichloroacetimidates. These materials are accessed directly from the hemiacetal, that is, from a free hydroxyl at the anomeric position, by treatment with trichloroacetonitrile and a suitable base. Depending on the type of base used either the α- or β-trichloroacetimidate may be obtained, as desired, as follows

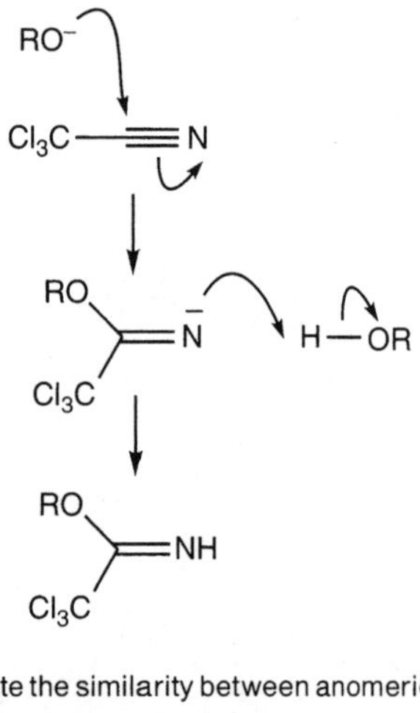

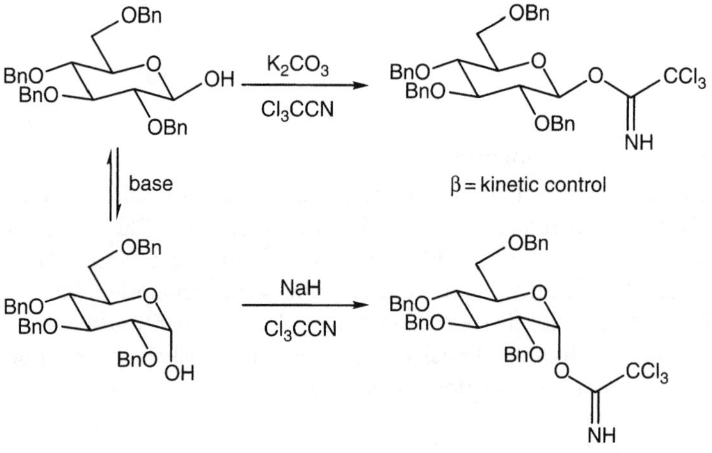

β = kinetic control

α = thermodynamic control

Note the similarity between anomeric trichloroacetimidate formation and the acetylation reactions documented in Chapter 4, wherein either the α- or β-anomeric acetates could be formed preferentially under different reaction conditions.

With a weak base such as potassium carbonate *mutarotation* occurs faster than the nucleophilic attack onto the trichlororacetonitrile by the anomeric hydroxyl, and so the β compound is formed predominantly; the equatorial β-anomeric OH is more nucleophilic than the axial α one. With a strong base such as NaH, deprotonation of the anomeric OH is effectively complete and alkylation probably occurs faster than mutarotation. However, NaH also catalyses a slow equilibration of the product α- and β-anomeric trichloroacetimidates and so the thermodynamically more stable α-isomer is finally formed exclusively.

Activation of anomeric trichloroacetimidates is usually achieved by treatment with a Lewis acid such as $BF_3 \cdot Et_2O$. Addition of an acceptor furnishes disaccharides in good yield, usually via an S_N2 type reaction with inversion of configuration at the anomeric centre, although, as usual, when there are ester protecting groups at the 2-position of the glycosyl donor, neighbouring group participation and an S_N1 type mechanism operates.

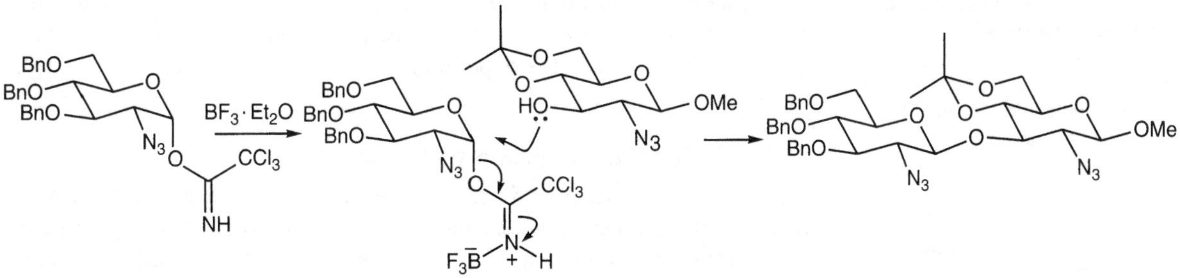

7.3.6 Glycals

The final class of glycosyl donors that we should mention are those derived from glycals. We saw in Chapter 5 how glycals can be readily synthesised by zinc mediated reduction of glycosyl bromides, which occurs with concomitant elimination of the protected 2-hydroxyl group. Epoxidation of the so-formed glycal, which is most easily performed with dimethyldioxirane, **7.7**, occurs stereoselectively on the less hindered bottom face to produce an epoxide such as **7.8**, termed a 1,2-anhydrosugar. These epoxides readily act as glycosyl donors when treated with suitable glycosyl acceptors under Lewis acid catalysis. Nucleophilic attack and ring-opening occur in an S_N2 fashion, leading to the stereoselective formation of 1,2-*trans* glycosides, namely β-glucosides, for example **7.9**.

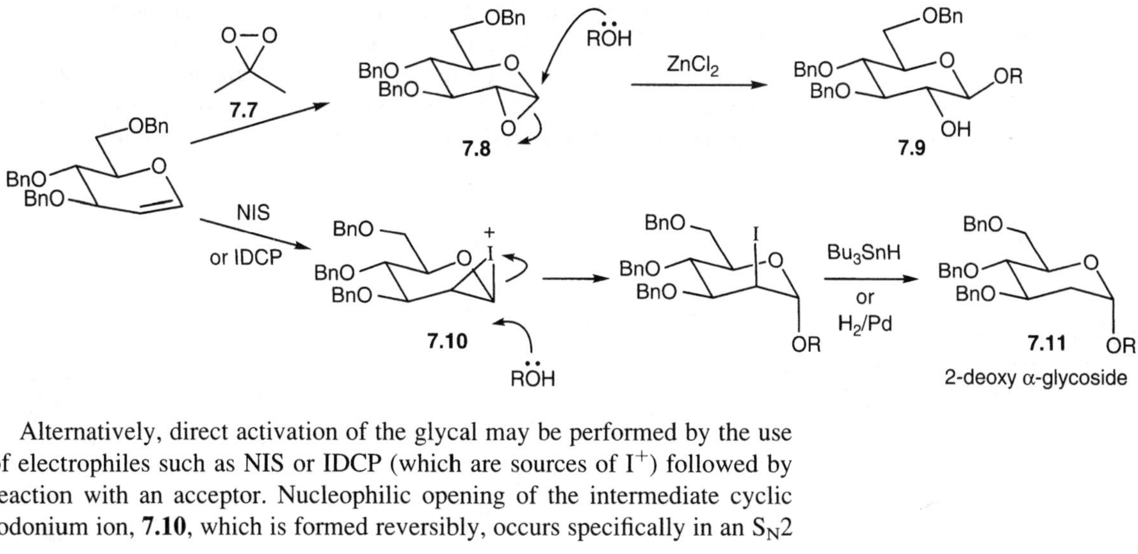

Alternatively, direct activation of the glycal may be performed by the use of electrophiles such as NIS or IDCP (which are sources of I^+) followed by reaction with an acceptor. Nucleophilic opening of the intermediate cyclic iodonium ion, **7.10**, which is formed reversibly, occurs specifically in an S_N2

fashion with diaxial opening under stereoelectronic control, to produce the α-manno 2-iodoglycoside. Simple removal of iodine by reduction, either using free radical methods (Bu$_3$SnH) or catalytic hydrogenation, leads to the synthesis of α-2-deoxy glycosides such as **7.11**.

7.4   Control of stereochemistry

As noted previously, the stereochemical outcome of glycosylation is one of the hardest elements to control. Obviously, for an efficient synthesis, one ideally requires the production of a single anomer in a predictable sense. Many factors influence the stereochemical outcome of glycosylation reactions and we will have the opportunity to review a few of them, in passing, below.

7.4.1   Neighbouring group participation

As we saw both in Chapter 5 and earlier on in this chapter, the stereochemical outcome of a glycosylation reaction, in which the glycosyl donor possesses an ester protecting group, on the 2-hydroxyl invariably proceeds via neighbouring group participation to yield exclusively the 1,2-*trans* glycoside product. Nucleophilic attack by the hydroxyl of the glycosyl acceptor on the cyclic oxonium ion occurs in an S$_N$2 fashion with the corresponding inversion of configuration. This, therefore, means that if we require the synthesis of a 1,2-*cis* linkage then we must have a non-participating group on the 2-hydroxyl of the glycosyl donor.

7.4.2   Solvent

The reaction solvent can have a marked effect on the stereochemical outcome of the glycosylation reaction. A good example of this is the stereoselective formation of β-linked disaccharides from glycosyl donors that do not possess a participating group at the 2-position, through the use of acetonitrile as the solvent. The reaction is thought to proceed via an S$_N$1 pathway to first give an intermediate glycosyl cation, which then forms an α-nitrilium ion by co-ordination to a solvent molecule. In the following example, trimethylsilyl triflate (TMSOTf) is used as the Lewis acid activator for the anomeric trichloroacetimidate, which presumably forms a glycosyl cation which is then trapped by the solvent to form the anomeric nitrilium ion. The glycosyl acceptor then attacks this α-nitrilium ion in an S$_N$2 reaction, with inversion of configuration at the anomeric centre, to produce, almost exclusively, the β product.

The distinction between 1,2-*trans* and 1,2-*cis* glycosides is a useful one to make. Since the former may be readily made by taking advantage of neighbouring group participation, they are considerably easier types of linkages to synthesise.

β-glucoside   α-mannoside

1,2-*trans* glycosides = easy to make

α-glucoside   β-mannoside

1,2-*cis* glycosides = hard to make

24:1 β : α

7.4.3 Molecular tethering—intramolecular aglycon delivery (IAD)

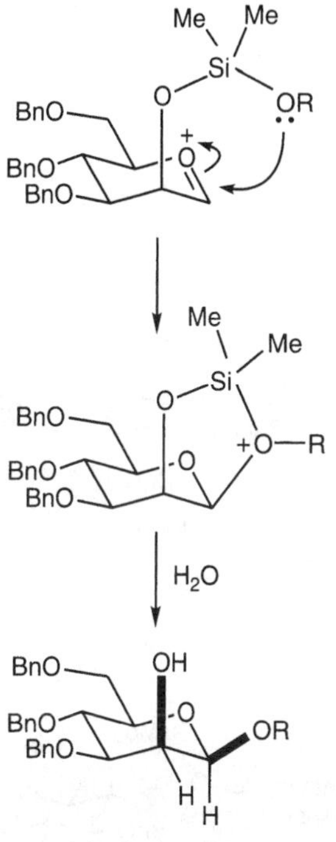

Since 1,2-*trans* glycosidic linkages may be stereoselectively formed by the use of ester-type, participating protecting groups on the 2-hydroxyl group of the glycosyl donor, the question therefore arises as to how we can stereoselectively form 1,2-*cis* glycosidic linkages? Although there are several potential solutions, a particularly elegant way to overcome this synthetic problem is to temporarily tie together the donor and acceptor before the glycosylation reaction is performed, in order to make glycosylation intramolecular and, hopefully, in an orientation that will favour the desired stereoselectivity.

One of the first examples of this type of approach was developed by Stork and co-workers for the formation of the difficult β-mannosyl linkage. They temporarily tied the glycosyl acceptor to the 2-hydroxyl group of the glycosyl donor via a silicon ether tether:

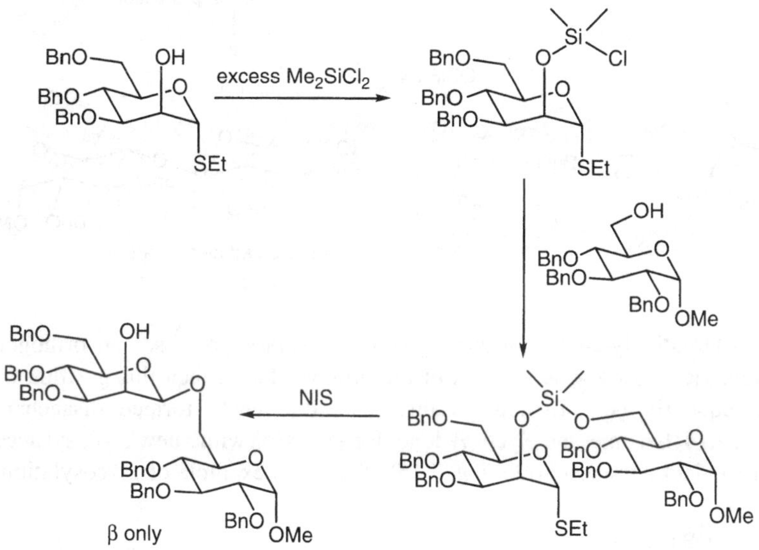

1,2-*cis* glycoside
a β-mannoside

Subsequent activation of the donor thioglycoside was followed by intramolecular delivery of the acceptor to the top (β) face of the mannose glycosyl donor. The fact that the glycosylation reaction proceeds via this intramolecular pathway, where the relative orientations of the donor and acceptor are fixed, ensures that the newly formed linkage is formed stereoselectively, and in this case has to be β. Similar approaches have been used for the stereoselective formation of α-*gluco* and α-*galacto* disaccharides.

7.5 Oligosaccharide synthesis

So far we have merely considered the formation of a single glycosidic linkage to allow disaccharide formation. However, the vast majority of interesting, biologically relevant carbohydrates are larger oligosaccharides. Obviously, the formation of subsequent glycosidic bonds can be performed on an iterative basis. This can be accomplished in two separate directions. If the original glycosyl donor used to form the disaccharide possessed a suitable protecting group pattern, then one hydroxyl of the disaccharide formed by this first glycosylation (e.g. **7.12**) could be selectively deprotected. In this way, the newly formed disaccharide could then act as an acceptor in a subsequent glycosylation reaction, with a new donor, to yield a trisaccharide (e.g. **7.13**). The following scheme illustrates an example of this glycosylation direction, which is known as **extension of the non-reducing end**.

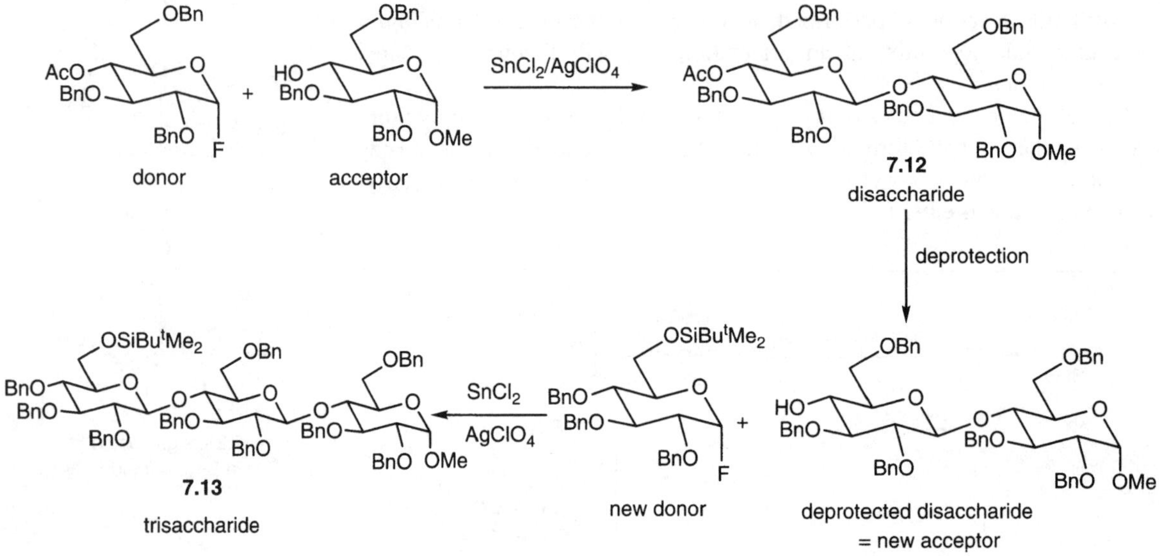

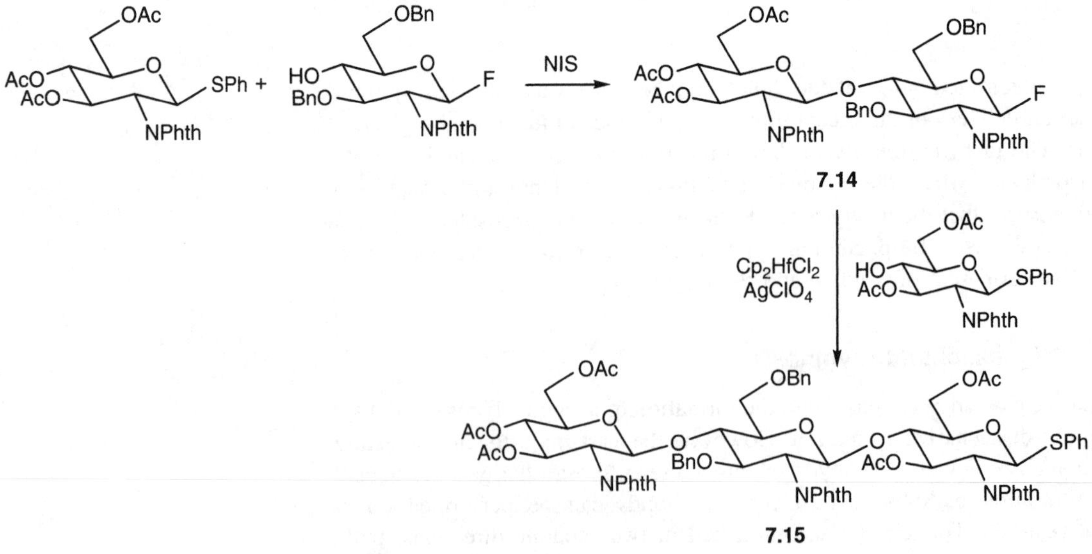

Alternatively, if the original glycosyl acceptor possesses an orthogonal anomeric leaving group to that of the original donor, then this position may subsequently be activated. In this case, the newly formed disaccharide (e.g. **7.14**) becomes the glycosyl donor for a reaction with a new glycosyl acceptor to yield a trisaccharide (e.g. **7.15**). This is an example of glycosylation in the opposite direction, which is known as **extension of the reducing end**. It is clear that simple iteration of both of these types of processes can quite quickly allow access to large oligosaccharide structures, albeit in a linear synthetic fashion. There are also many other ingenious approaches to the rapid assembly of oligosaccharide structures which, unfortunately, we do not have the opportunity to discuss; once again the interested reader is encouraged to consult the suggestions for further reading.

7.6 Solid phase synthesis

Solid phase synthesis is a technique whereby a compound of interest is initially attached to an insoluble support, often a polymer, via a linker which is cleavable under a specific set of conditions. A sequence of synthetic steps is then performed on this material as for 'normal' solution phase synthesis (where the molecule of interest is dissolved in a solvent), but with the added advantage that purification and handling of the reaction products are greatly simplified. For example, purification and isolation after each reaction step can simply consist of filtration to retrieve the immobilised material minus by-products. At the end of the desired synthetic sequence the linker is cleaved and the compound of interest is released from the solid support.

The use of solid phase synthesis has revolutionised the synthesis of oligonucleotides and peptides. Basically, almost any sequence of these two bio-oligomers is now readily available upon demand by a completely automated process. The development of similarly efficient solid phase methodology for the synthesis of oligosaccharides would be revolutionary. However, the extra complications involved in the synthesis of oligosaccharides, namely those of regioselectivity and stereoselectivity, are still confounding organic chemists. Since solid phase synthesis requires highly efficient reactions that give single products in near quantitative yields, and glycosylation reactions often still do not fulfil these criteria, we can conclude that there is still much work to be done by the carbohydrate chemist!

Phthalimides are commonly used as protecting groups for nitrogen in carbohydrate chemistry. They perform neighbouring group participation, hence facilitate the formation of β *gluco* linkages, but unlike acetate they have little tendency to migrate.

Although machines have now been built, these are, unfortunately, not general in the range of oligosaccharides that they can make. They synthesise a particular bond type well but others poorly. Given the enormous permutations of oligosaccharide synthesis this would require, 1×10^{12} machines, to make, for example, all possible hexasaccharides, and a pretty big laboratory! The real challenge is to find an adaptable method for all oligosaccharides and this remains one of organic chemistry's great unsolved problems.

7.7 Summary

At the end of this chapter you should be able:

- to define the terms **glycosylation reaction**, **glycosyl donor** and **glycosyl acceptor**;
- to understand the problems of regio- and stereochemistry inherent in the synthesis of di- and oligosaccharides;
- to recall how to synthesise and activate the most commonly used classes of glycosyl donors;
- to describe how 1,2-*trans* glycosidic linkages can be made by taking advantage of neighbouring group participation;
- to understand the importance of solvent in determining the stereochemical outcome of glycosylation reactions;
- to describe how 1,2-*cis* glycosidic linkages can be made by temporary molecular tethering (or IAD, intramolecular aglycon delivery);
- to understand the principles of iterative oligosaccharide synthesis and the use of orthogonal glycosyl donors/acceptors.

7.8   Questions

1. Predict the products of the following reactions.

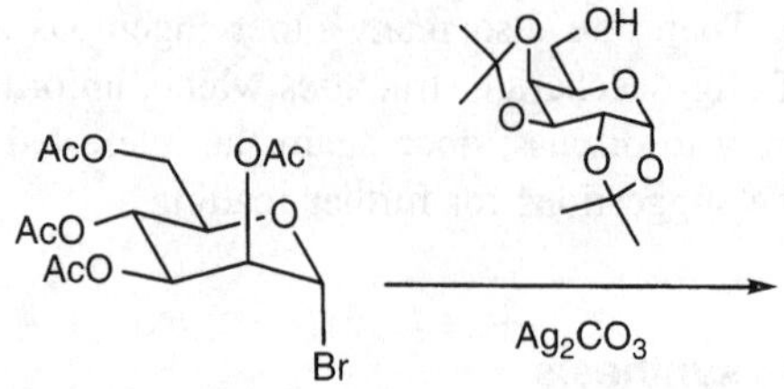

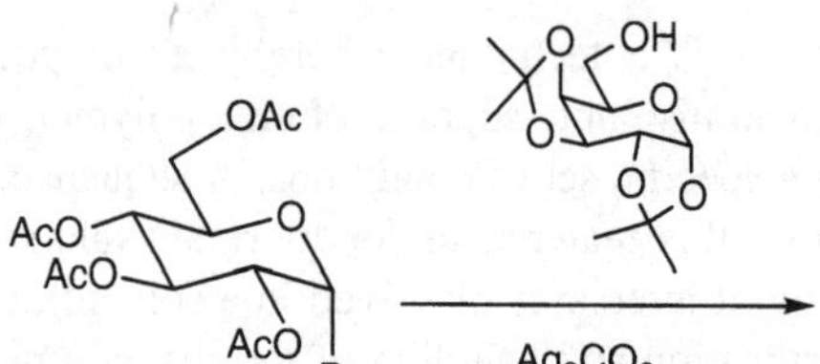

2. Explain how you would achieve the following overall transformation (more than one step may be required).

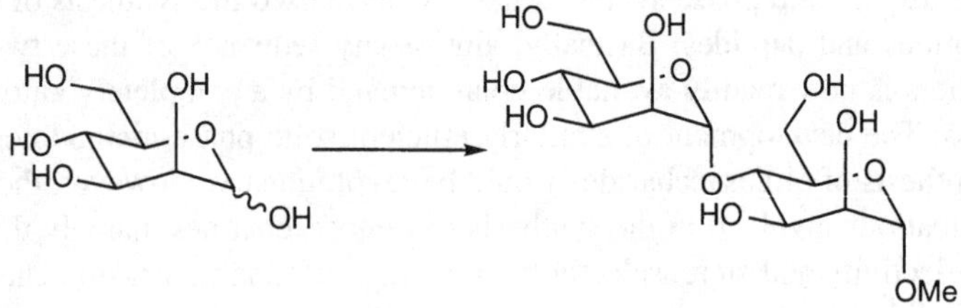

3. Side reactions can be a problem in glycosylation reactions using **7.16**. Through mechanistic reasoning, work out which side reaction found when using **7.16** as a glycosyl donor is suppressed by the use of **7.17**? (Hint: you met this reaction in Chapter 5.)

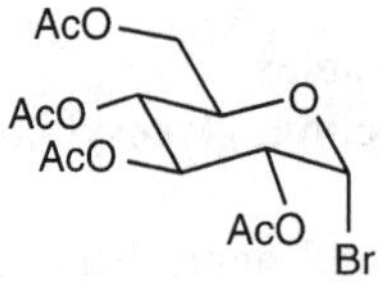

7.16 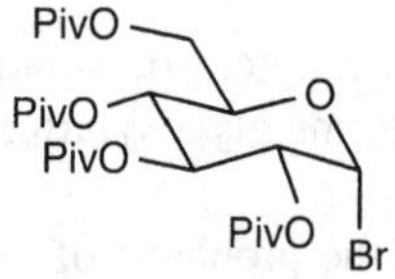   **7.17**

8 Enzymatic disaccharide formation

8.1 Tackling the problems of chemical disaccharide formation

As described in the previous chapter, the chemical formation of a particular glycosidic bond between, for example, two monosaccharides involves the consideration of three factors:

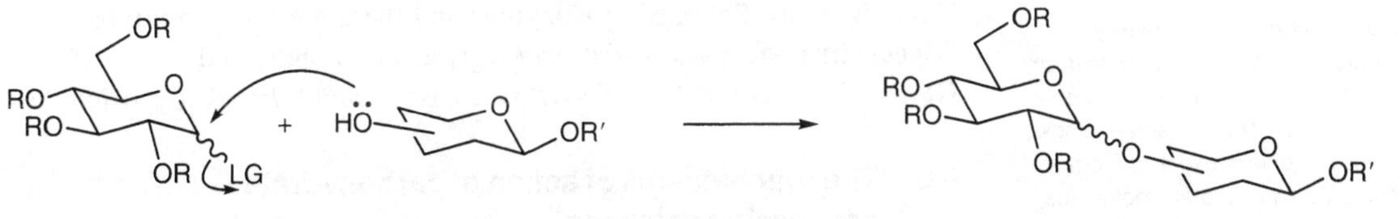

| glycosyl donor | glycosyl acceptor | disaccharide |

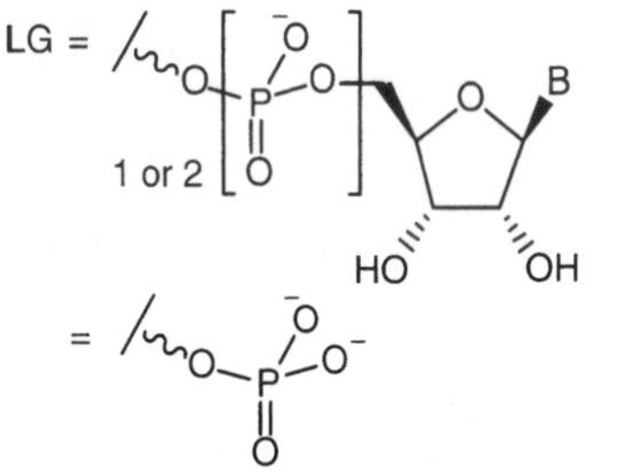

1. Reactivity: First, the glycosyl donor needs to be reactive enough to form the bond, that is, it must have a good leaving group at the anomeric centre.

2. Regioselectivity: We need to ensure that only one of the hydroxyl groups in the glycosyl acceptor acts as a nucleophile, otherwise, a mixture of disaccharides will result. Furthermore, we need to prevent the hydroxyl groups in the glycosyl donor from reacting with another molecule of itself to form oligomers of repeating glycosyl donor units.

   These requirements usually necessitate a number of reactions to protect both the donor and acceptor molecules so that unwanted bonds are not formed. For disaccharide formation this usually requires the protection of all of the hydroxyl groups in the glycosyl donor and all except one (the one that is destined to form the glycosidic bond) in the glycosyl acceptor. This often involves a number of difficult protection steps before we can couple donor to acceptor. Of course, these groups also have to be removed at the end of the synthetic sequence, and this requires yet more steps. Clearly, the complexity of these additional steps is increased further when oligosaccharides containing more saccharide units are to be coupled.

3. Stereoselectivity: By forming the glycosidic bond, we create a stereogenic centre at the anomeric carbon of what was the glycosyl donor unit. How do we control its configuration to form either α- or β-linked structures stereoselectively?

Despite the development of a number of chemical techniques that allow partial control of the configuration at the anomeric centre, these methods are

In nature

It has been estimated that on average, for each glycosidic bond formed in chemical oligosaccharide synthesis, six additional associated protection and deprotection steps are required.

rarely general. As a result, mixtures of products are often formed that require careful purification. The formation of side products with the wrong stereochemistry also lowers the yields of the desired product.

8.2   Glycosyltransferases and glycosidases

Enzymes are often treated only as biological reagents but they are really no different to any other catalysts used in chemistry.

Some advantages are: **specificity**—they usually do one type of reaction very well; **no need for protecting groups**—only particular groups will react, even in the presence of others; **efficiency**; **mild conditions**; **environmentally friendly**—they are non-toxic & biodegradable.

Disadvantages include: **overly specific**—it may be hard to use them as catalysts for a range of reactions; **careful conditions**—in some cases precise conditions of pH, cofactors (associated non-protein molecules that are required by some enzymes for activity), and temperature may be needed.

In Chapter 7 we saw a number of chemical approaches to the control of the factors surrounding glycosylation. We can solve many of the problems by taking a leaf out of Nature's book. The existence of a vast number of different naturally occurring oligosaccharides requires that biological systems solve these problems too. Nature, however, does so in one step and without the use of protecting groups. By examining the way in which Nature constructs and dismantles oligosaccharides we can learn lessons that allow us to synthesise oligosaccharides more efficiently. Evolutionary changes over many millions of years have led to the existence of families of enzymes which are powerful catalysts that are perfectly adapted to these functions. These are termed **Carbohydrate Processing Enzymes** and there are two main types: **Glycosyltransferases**, which *make* glycosidic bonds; and **Glycosidases** (short for glycosylhydrolases) which *break* glycosidic bonds.

8.3   The mechanisms of action of carbohydrate processing enzymes

Despite their opposing functions there are similarities in the way that glycosyltransferases and glycosidases catalyse reactions. Making comparisons allow us to see how we can use these enzymes in chemically similar ways to form glycosidic bonds and show how their use can solve some of the problems we outlined above.

Before we consider these reactions in detail, it is helpful to consider the region of these enzymes that controls the reactions they catalyse. This is called the active site, and can be considered as being made up of two parts: the catalytic site, which increases the *rate* of the reaction, and the binding site or sites which hold(s) the reactants in place and determines the *selectivity* of the reaction.

Let us consider these in terms of the problems of glycosidic bond formation that were outlined above:

Here we have chosen an S_N1-type mechanism to illustrate this discussion of reactivity but, just as for chemical glycosylation, enzymatic glycosylation may be S_N1 or S_N2.

1. Reactivity—Just as for most chemical methods of forming glycosidic bonds, enzymatic mechanisms also rely on the loss of a good leaving group from the anomeric centre of the glycosyl donor (Fig. 8.1). This bond breaking is typically assisted by the involvement of the lone pair on the ring oxygen atom and leads to the formation of a glycosyl cation intermediate. The transition state that leads to the formation of this reactive intermediate is stabilised by the catalytic site of the enzyme. As the formation of this transition state is the slowest step, lowering the associated energetic barrier causes an increase in the rate of reaction. The shape of the binding site of the enzyme now holds this reactive species ready for attack by a nucleophile such as the hydroxyl group of another sugar, the glycosyl acceptor.

2. Regioselectivity—Because the reactants fit perfectly into the binding sites of the enzyme in only one particular orientation, the shape of the

enzyme determines which groups will react. Both the glycosyl acceptor and glycosyl donor are held tightly in the enzyme's binding site. Their relative orientations allow the reaction of only one of the hydroxyl

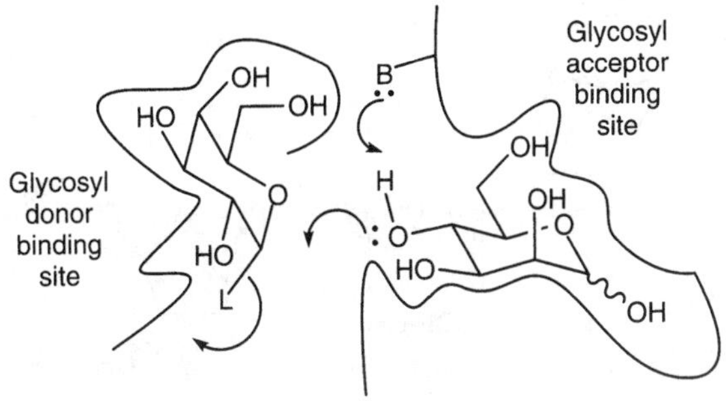

Fig. 8.1

groups of the acceptor. The other hydroxyl groups of the glycosyl acceptor are buried in the folds of the binding site and they are therefore temporarily protected and so do not react. Furthermore, the reactive anomeric carbon atom in the glycosyl donor is close enough to only one particular hydroxyl group to react.

3. Stereoselectivity—The enzyme also binds the reactive glycosyl donor in such a way that it can only be attacked from one side: either above or below. The result is that the formation of only one stereochemical configuration is possible. Which configuration is formed therefore depends on the shape of the enzyme active site and different enzymes have evolved with differently shaped active sites; some allow only the formation of α products whilst some only β. Let us consider the two types of carbohydrate processing enzymes in more detail.

8.4 Glycosyltransferases

Glycosyltransferases (Gly-Ts) make glycosidic bonds using the same basic approach that has been developed by chemists. They bring together a glycosyl acceptor hydroxyl group to react with a glycosyl donor bearing a good leaving group.

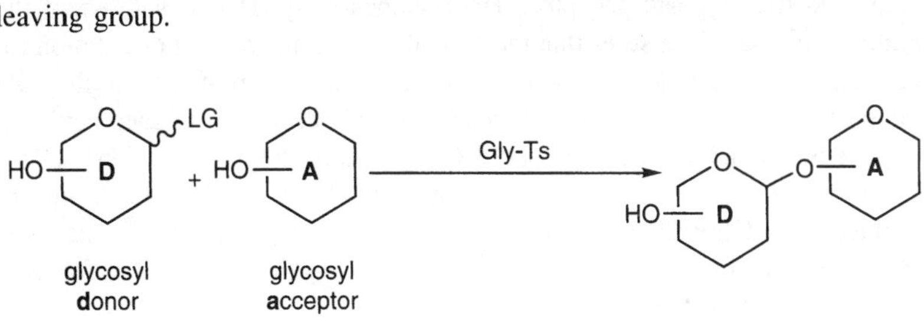

For glycosyltransferases the glycosyl donor is usually a glycosyl nucleotidediphosphate. Despite its apparently complex structure, the nucleotide diphosphate part is really just Nature's equivalent to leaving groups such as bromide and chloride that were described in Chapter 5. In fact, under suitable

The names of enzymes describe the reactions that they catalyse. *Glycosyl* **transfer**ases **transfer** *glycosyl* units whilst *glycosyl* **hydrol**ases **hydrol**yse the *glycosyl* bond. We can be more specific about the reaction by simply expanding the name. For example, the enzyme **mannose β(1,4)-galactosyltransferase** transfers a **galactosyl** unit to the **O-4** atom of **mannose** with the formation of a β-anomeric bond. Therefore, the name of the enzyme catalyst allows us to determine the specific product, which in this case is:

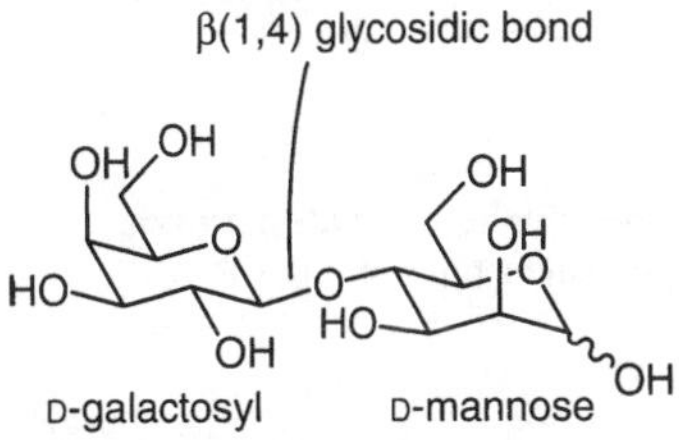

4-*O*-(β-D-galactopyranosyl)-D-mannopyranose

Nucleosides are molecules in which a purine (adenine (A), guanine (G)) or pyrimidine (cytosine (C), thymine (T), uracil (U)) is linked to the anomeric centre of a pentose sugar (usually D-ribose) through a nitrogen atom. Addition of a phosphate group (mono-, di-, or tri-) to the OH-5 group of the pentose creates a nucleotide e.g.

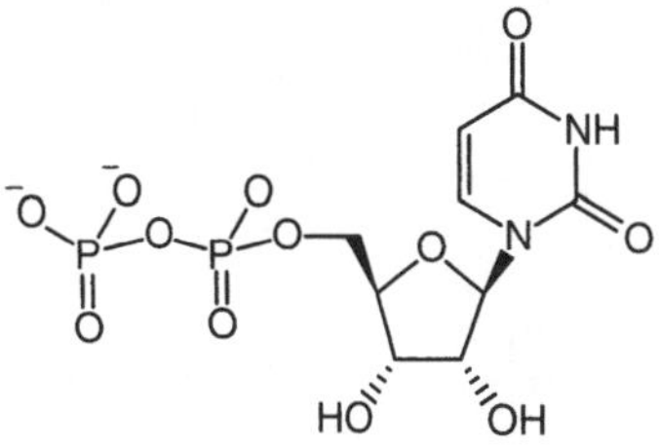

uridine diphosphate—a nucleotide

The types of glycosyltransferase have been further divided according to the nature of the leaving group in the glycosyl donor. There are two: *Leloir* type use glycosyl donors with nucleotide mono- or di-phosphate leaving groups at the anomeric centre **8.3**, *non-Leloir* type use other leaving groups such as phosphate groups **8.4**.

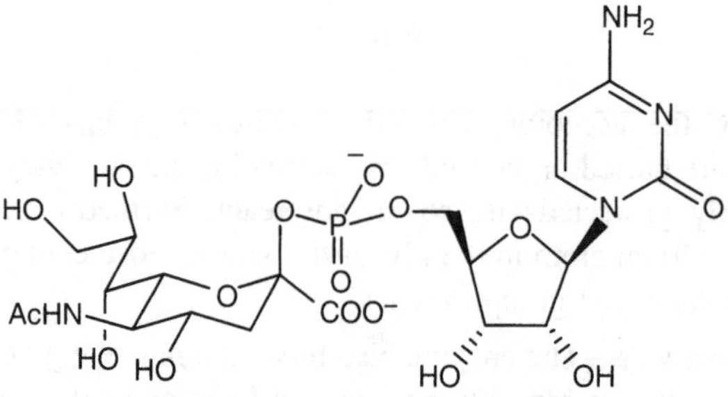

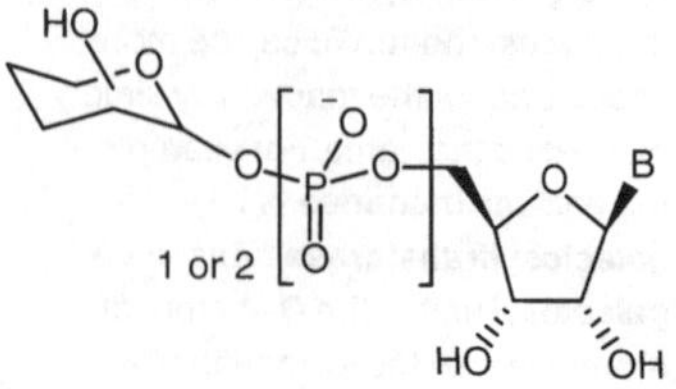

conditions, glycosyl nucleotidediphosphates can also be used as chemical glycosyl donors.

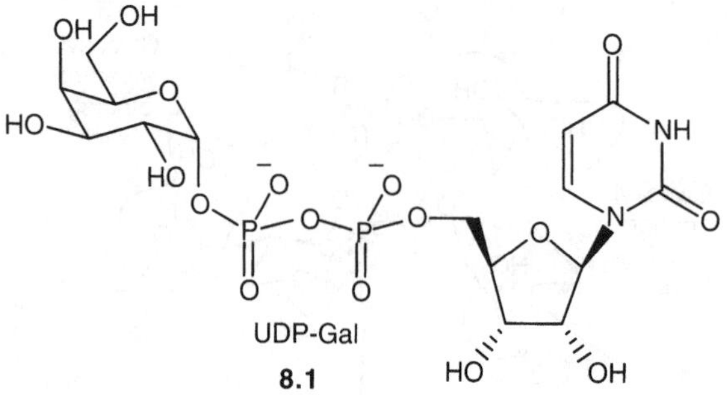

8.3

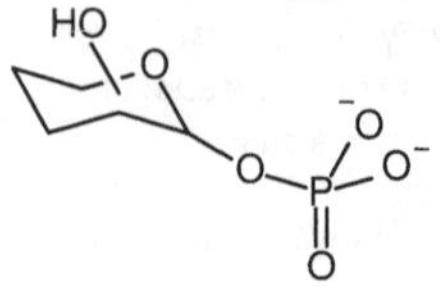

8.4

where B = appropriate purine or pyrimidine base of A, C, T, G, U.

UDP-Gal
8.1

CMP-sialic acid
8.2

The complexity of this leaving group portion also gives glycosyltransferases another level of selectivity. Particular enzymes not only bind the sugar portion of the glycosyl donor tightly but the nucleotide portion as well. As a result, glycosyltransferases only use certain corresponding pairs of sugar and nucleotide. For example, galactosyltransferases only bind uridine-5′-diphosphogalactose (UDP-Gal) **8.1** and not GDP or CDP, whereas sialyltransferases only bind sialic acid donors **8.2** that contain cytidinemonophosphate (CMP).

The glycosyltransferase in question binds the two reactants: the glycosyl donor and the glycosyl acceptor. Their shapes in space match those of the binding site and this ensures that they are aligned in the perfect orientation for reaction. For example, in mannose $\beta(1,4)$-galactosyltransferase, the OH group on carbon atom 4 of the mannose glycosyl acceptor is aligned perfectly for backside attack on the anomeric carbon atom of the glycosyl donor UDP-Gal **8.1**. Thus, only this hydroxyl group attacks in an S_N2 reaction and displaces the GDP leaving group. As for all S_N2 reactions this happens with inversion of configuration such that an α-galactosyl donor forms a β-galactoside (Fig. 8.2).

Each enzyme is so specific we can put many glycosyltransferases together to transfer several monosaccharides, one after another, in one reaction vessel. Since these enzymes have evolved to catalyse just one reaction they do not interfere with the other reactions even though they are present in the same solution.

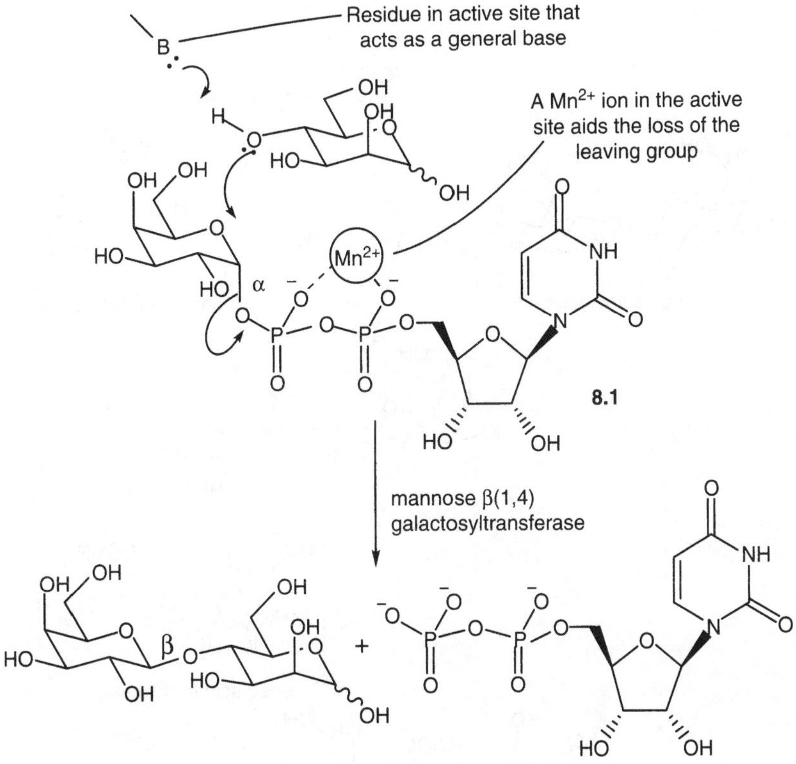

Fig. 8.2

A very good example of a sequential one-pot synthesis is the synthesis of the important tetrasaccharide sialyl Lewis-x (sLe^x) **8.5** shown in Fig. 8.3. In effect this an enzymatic factory for making complex oligosaccharides. We feed into the system simple unprotected sugars and derivatives as starting materials and sLe^x **8.5** is the product. Using protecting groups and standard chemical techniques the equivalent chemical synthesis took 31 steps!

As we have seen, the power of enzymes lies in their specificity. Glycosyltransferases are some of the most specific enzymes known. The shape selectivity imposed by their binding pockets is one of their principal advantages. Three perfectly shaped binding sites ensure that they only catalyse a reaction between the correct reactants to form a single particular bond type, that is, each will only bind a particular glycosyl acceptor to react with a particular glycosyl donor bearing only one type of nucleotidediphosphate. This specificity means that an enormous number of different species of glycosyltransferases exist—each one designed to make just one type of bond. Therefore, in theory, perfect catalysts are available for making all the varied glycosidic bonds found in Nature.

In reality, only a small fraction of the glycosyltransferases in Nature has been isolated. Isolation of new glycosyltransferases is sometimes difficult as they may be unstable and are found in only small quantities. Their intolerance of substrates that are not their preferred ones also leaves them with little or

We have only talked here about inverting glycosyltransferases but as for glycosidases (see Section 8.5) there are both inverting *and* retaining glycosyltransferases.

You will learn more about the important biological activities of sLe^x in Chapter 9.

The glycosyltransferases most frequently used in synthesis are:

β(1,4)galactosyltransferases;
α(2,3)sialyltransferases;
α(2,6)sialyltransferases;
α(1,2)fucosyltransferases;
α(1,3)fucosyltransferases;
α(1,6)fucosyltransferases;
N-Acglucosaminyltransferases.

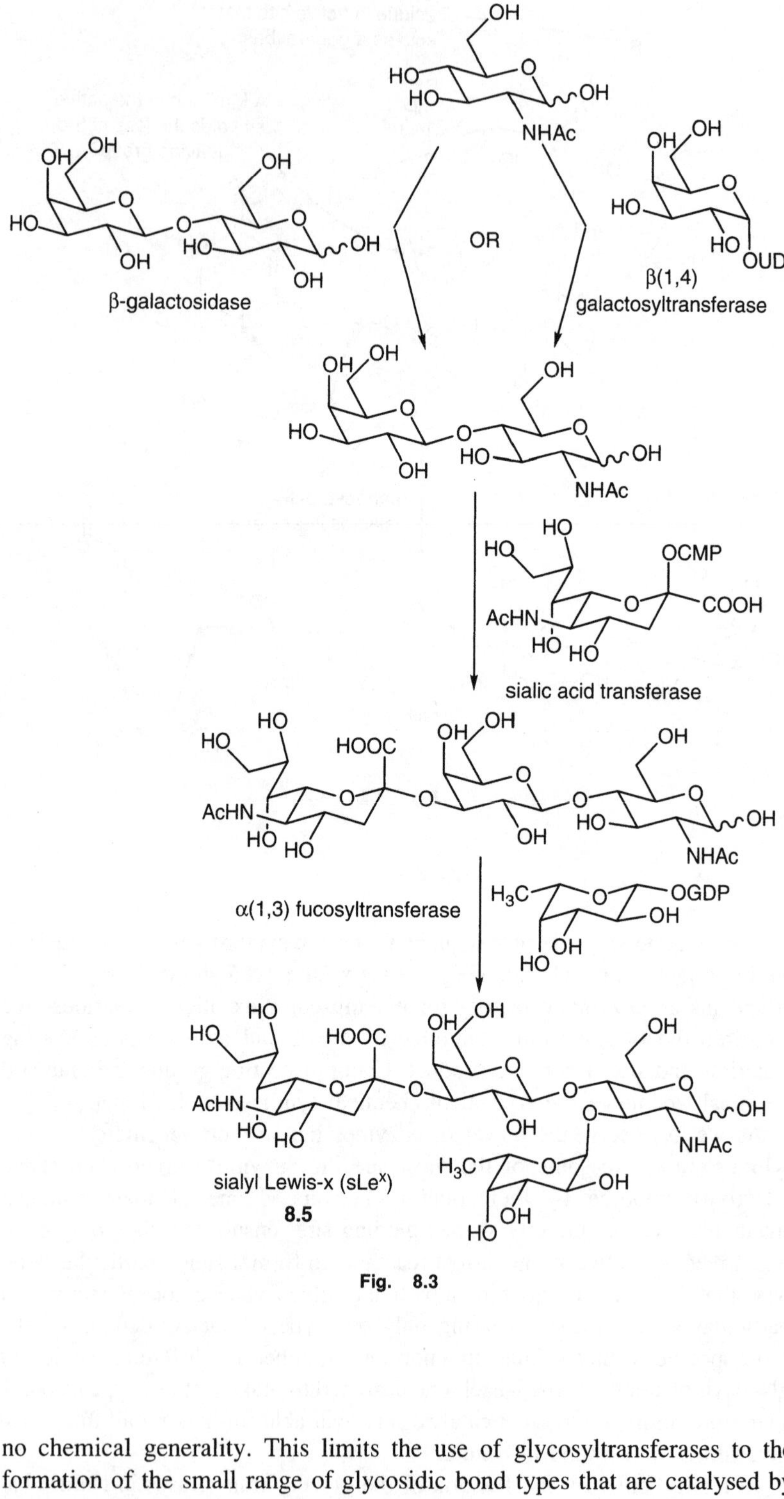

Fig. 8.3

no chemical generality. This limits the use of glycosyltransferases to the formation of the small range of glycosidic bond types that are catalysed by readily available enzymes. Furthermore, the nucleotidediphosphate glycosyl donors are hard to synthesise, although, as Fig. 8.4 shows, other enzymes can be used to overcome this problem.

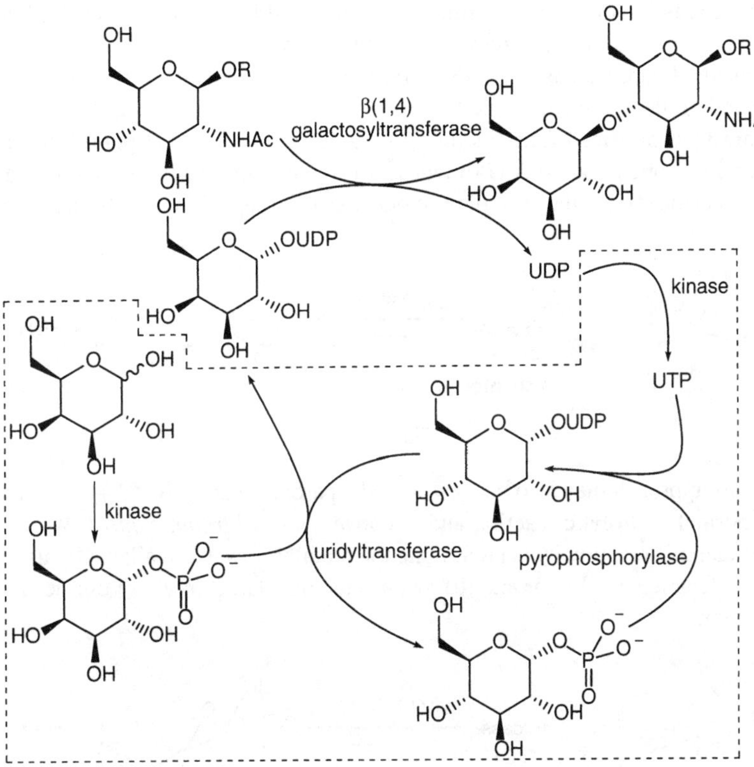

Fig. 8.4

Multi-enzyme systems have been developed that allow the sugar nucleotide glycosyl donor to be recycled *in situ* simply by adding enzymes. For example, kinases, pyrophosphorylases (which make phosphoester bonds) and nucleotidyltransferases will synthesise these donors.

Thus, the reaction of the galactosyltransferase in Fig. 8.4 is supported by a system of three linked enzyme-catalysed reactions in the dotted box that recycle UDP to Gal-UDP.

8.5 Glycosidases

At first inspection it may seem odd to use this type of enzyme, which has evolved to *break* glycosidic bonds (Fig. 8.5), to *make* them instead. However, if we look more closely at the mechanism of these enzymes we can see how such a process may work.

Fig. 8.5

Recall that acetal hydrolysis and formation proceeds via an S_N1-type mechanism and is a recurring theme in carbohydrate chemistry. Make sure you are familiar with the mechanism of this reversible reaction.

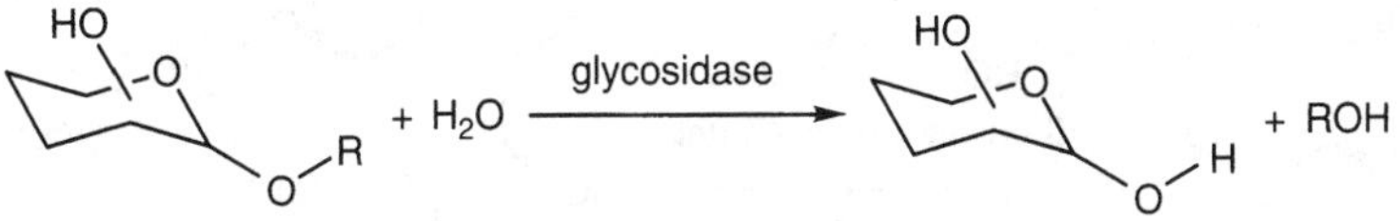

Typically, a glycosidase has carboxylic acid residues in its catalytic site. One acts as a general acid catalyst. More specifically, it acts by protonating the oxygen atom at the anomeric position of a glycosidic substrate to form an oxonium ion **8.6**. This converts the alcohol found at the anomeric centre into a good leaving group and so catalyses the formation of a glycosyl cation intermediate **8.7**. This closely resembles the chemical hydrolysis of glycosides described in Chapter 3.

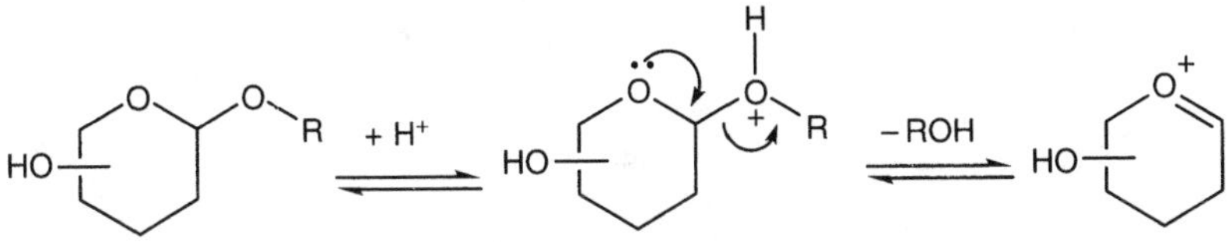

8.6　　　　　　**8.7**

Normally, this reactive intermediate is attacked by water as a nucleophile which leads to the reaction found in biological systems—the hydrolysis of the glycosidic bond. But, as we can see from Fig. 8.6, if we can replace the water by another alcohol nucleophile R'OH, such as the hydroxyl group of another sugar, we can instead form another glycosidic bond. This 'swapping' of one glycosidic bond for another is known as *transglycosylation*, in the same way as transesterification is the swapping of one alkoxy group of an ester for another.

R' in the glycosyl acceptor may be H, a simple alcohol or another sugar.

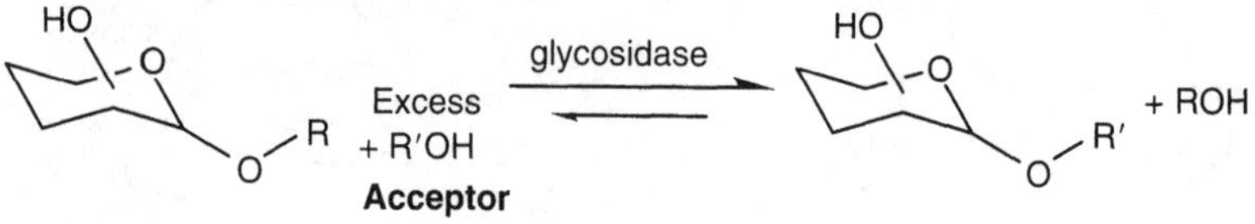

Fig. 8.6

This approach may also be used with the parent sugars (Fig. 8.7) to reverse the normal hydrolytic reaction and is known as *equilibrium control*. We can facilitate this reaction by (a) removing the amount of water available by using water-depleted conditions and (b) by introducing a large excess of our desired acceptor.

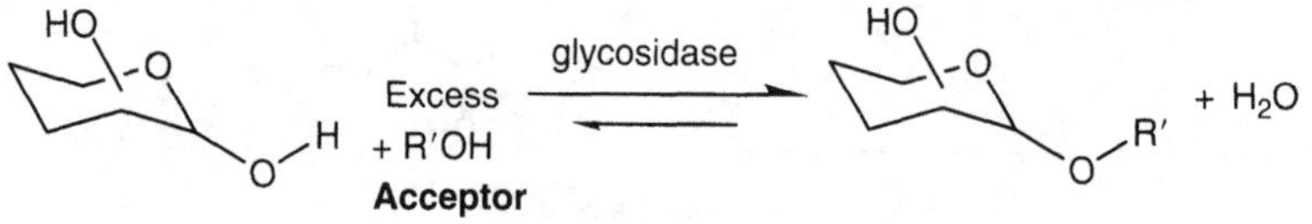

Fig. 8.7

Some examples of the varied reactions that can be carried out using, for example, β-galactosidase, in this way are shown in Fig. 8.8:

where R = H or

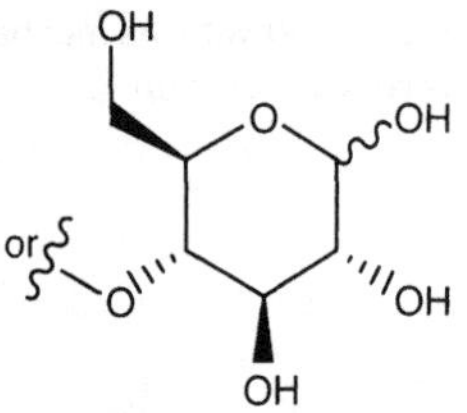

Note that because β-galactosidase only cleaves β-galactosides, then by the principle of microreversibility it only makes them too.

Also note that because of the specificity of β-galactosidase, if a disaccharide glycosyl donor containing β-galactose at its non-reducing end is used it does not react with the glucose portion and it may be treated as just another R group.

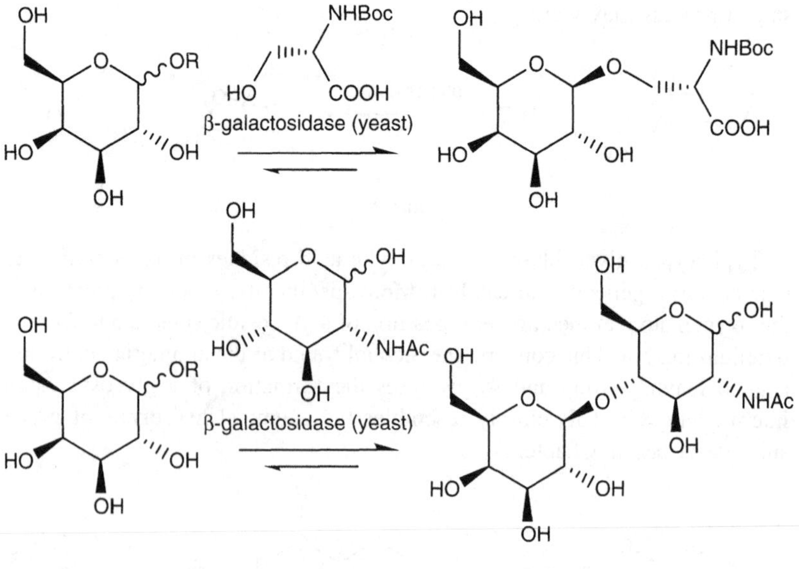

Fig. 8.8

We can also increase the rates of reactions that are catalysed by glycosidases by using other types of glycosyl donors with very good leaving groups. Two frequently used examples are glycosyl fluorides, such as **8.8**, or

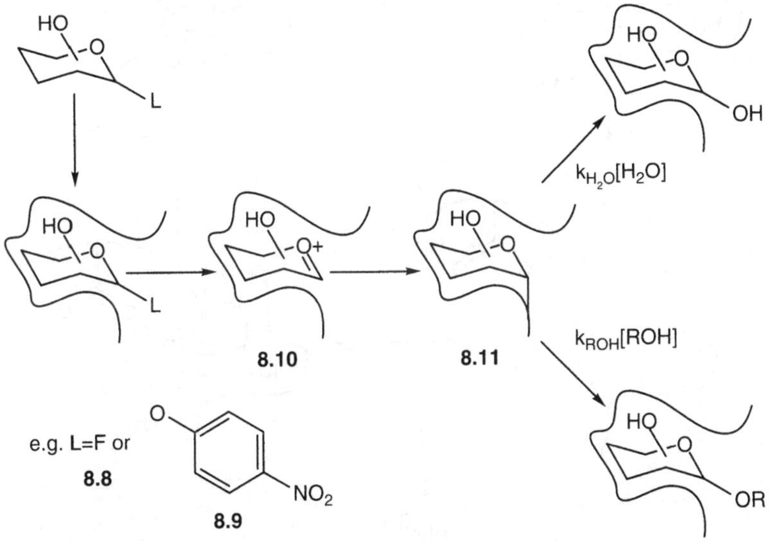

para-nitrophenyl glycosides, such as **8.9**. These glycosyl donors release poor nucleophiles that, unlike the alcohol leaving groups we have illustrated in Fig. 8.8, will not compete as nucleophiles. As a consequence, the reverse reaction is very slow and the reaction is now essentially irreversible. This means that we no longer need to use an excess of glycosyl acceptor to displace the equilibrium. The exclusion of water is still beneficial, however, as we are now involved in a competition reaction in which the glycosyl acceptor competes to react with either a glycosyl cation intermediate **8.10** or an enzyme-bound intermediate **8.11**. The success of the desired reaction will now be determined by the relative rates of reaction. We want $k_{H_2O}[H_2O]$ to be low compared to $k_{ROH}[ROH]$, where ROH is a general glycosyl alcohol acceptor. This approach is known as **kinetic control**.

Glycosidases can be divided into two types according to the stereospecificity they show in glycosidic bond formation: **retaining** and **inverting**. In *inverting* glycosidases (Fig. 8.9), as the name implies, the configuration of the anomeric centre in the product is inverted relative to that of the starting glycoside. Just as for glycosyltransferases this is ensured by the shape of the active site that binds the glycosyl donor. The glycosyl donor is shielded by the folds of the binding site such that the approach of the nucleophile that attacks it, for example, the hydroxyl group of the glycosyl acceptor, can only occur from the opposite face to the leaving group. Hence a β-galactoside **8.12** gives rise to an α-galactoside **8.13**.

Retaining glycosidases give rise to products with the same anomeric configuration as in the glycosyl donor through the double inversion mechanism shown in Fig. 8.10. A second carboxylic acid in the active site intercepts the glycosyl cation to form a glycosyl-enzyme intermediate **8.14**. This intermediate **8.14** is then attacked by the nucleophile from the *same* face from

A two step S_N1 process has been shown here for inverting glycosidases. An S_N2 process may also be possible.

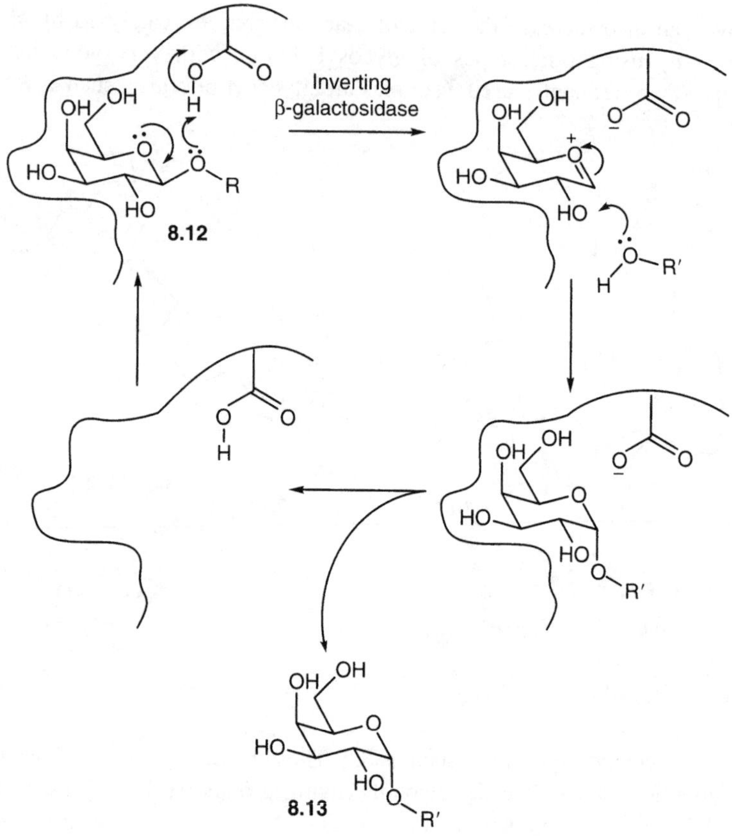

Fig. 8.9   Inverting glycosidase mechanism.

which the leaving group departed. Hence β-galactoside **8.15** gives rise to β-galactoside **8.16**.

Glycosidases are much less specific than glycosyltransferases as they only possess a tight binding site for the glycosyl donor. The glycosyl acceptor binding site is often less strict in its shape requirements and in some cases does not even exist. This can be viewed as an advantage as it allows a broader range of glycosyl acceptors to be used in the same enzymatic system. For example, as Fig. 8.8 illustrates, a β-galactosidase from yeast will accept very structurally diverse glycosyl acceptors such as *N*-acetylglucosamine or the hydroxyl in the side chain of the amino acid L-serine. Here the disadvantage is that some of the strict regioselectivity that we observed with glycosyltransferases may be lost in glycosidase-catalysed reactions. For example in Fig. 8.11, the *N*-acetylglucosamine derivative **8.17** reacts to give a mixture of all possible regioisomers **8.18**, **8.19**, and **8.20**.

Advantages of glycosidases are that they are typically more stable and more easily isolated than glycosyltransferases. Furthermore, the required glycosyl donors, which in the case of equilibrium controlled reactions can be simple glycosides or even the parent sugars themselves, are usually easy to make.

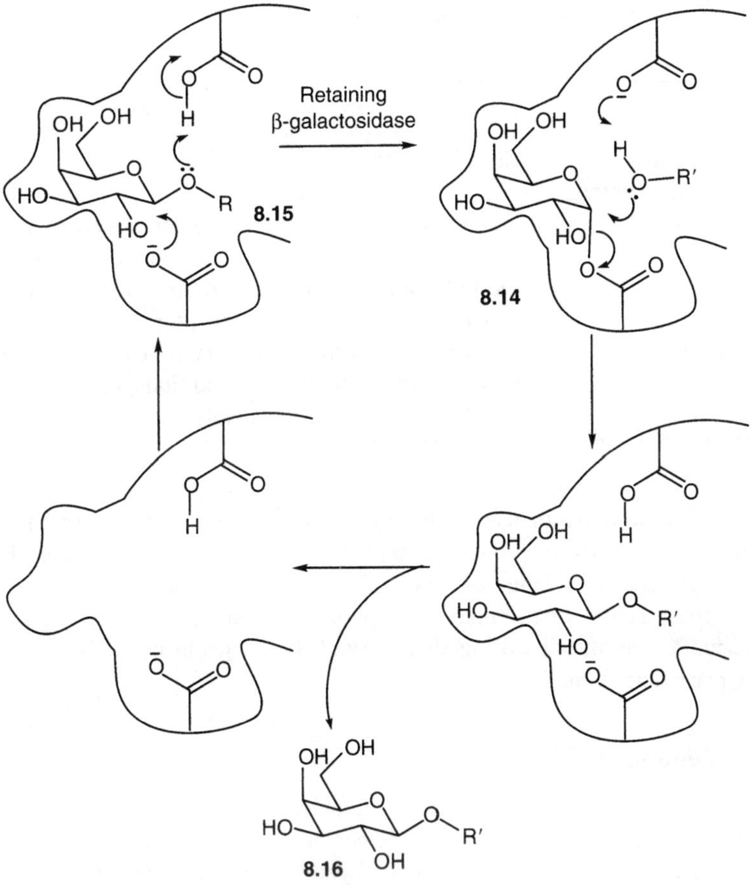

Fig. 8.10 Retaining glycosidase mechanism.

Note that in retaining glycosidases nucleophilic attack by the hydroxyl group of the acceptor is aided by the same carboxylate that originally acted as a general acid. Now it acts as a general base by deprotonating the attacking hydroxyl group and is returned to its original protonated form. The enzyme is thus unchanged overall by the reaction. Recall that this regeneration of original state is a fundamental part of the definition of a catalyst.

Enzymes which are of the same type but have been isolated from different organisms usually have different amino acid compositions and structures. This might also mean that they have different properties and selectivities. Therefore, strictly, when an enzyme is described, both the type e.g. β-*galactosidase*, and the source e.g. *from yeast* (or even more specifically the genus and species), should be quoted.

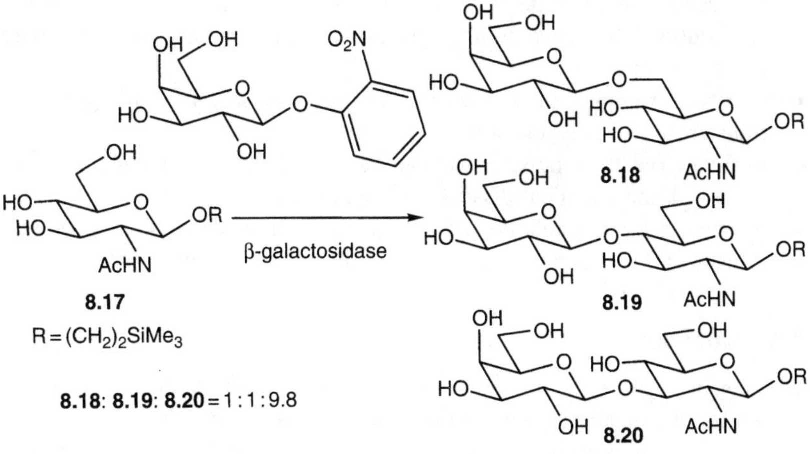

R = $(CH_2)_2SiMe_3$

8.18: **8.19**: **8.20** = 1 : 1 : 9.8

Fig. 8.11

The *ortho*-nitrophenyl glycoside shown here is, like *para*-nitrophenyl glycosides, also an activated donor with a good leaving group.

8.6   Glycosidases versus glycosyltransferases

	Glycosyltransferases	Glycosidases
Efficiency in disaccharide synthesis	Usually $> 80\%$	Often $\sim 40\%$
Specificity	For the glycosyl donor, glycosyl acceptor *and* the nucleotide leaving group	Only for the glycosyl donor and sometimes partially for the glycosyl acceptor
Ease of use	Need buffers, correct glycosyl donor, hard to isolate	Very robust, tolerant of conditions, easier to isolate

In deciding whether to use a glycosyltransferase or a glycosidase in disaccharide synthesis it should be remembered that they each have their advantages and disadvantages, some of which are summarised in the above table. These complement rather than oppose each other, and these two types of enzymes are often used together in synthetic routes to take advantage of their particular strengths.

8.7   Summary

At the end of this chapter you should be able:
- to realise the factors that need to be considered in disaccharide formation;
- to describe the two classes of carbohydrate processing enzymes;
- to understand how enzymes are able to be selective in the reactions that they catalyse;
- to describe the mechanism of action of glycosyltransferases including the types of leaving groups that they employ;
- to describe the mechanisms of action of glycosidases for both inverting and retaining classes;
- to recall and explain the two strategies that allow the use of glycosidases as catalysts for *synthesising* glycosidic bonds when naturally they *cleave* them;
- to give examples of syntheses that may be catalysed by glycosyltransferases and glycosidases;
- to summarise the relative advantages and disadvantages of using enzyme-catalysed and chemical disaccharide synthesis;
- to summarise the relative advantages and disadvantages of glycosyltransferase- and glycosidase-catalysed disaccharide syntheses.

8.8   Questions

1. Suggest syntheses of the following disaccharides by choosing appropriate chemically-, glycosyltransferase- or glycosidase-mediated methods. Take into account all the conditions that may be required and use them to help you make your choice of method.

(a)

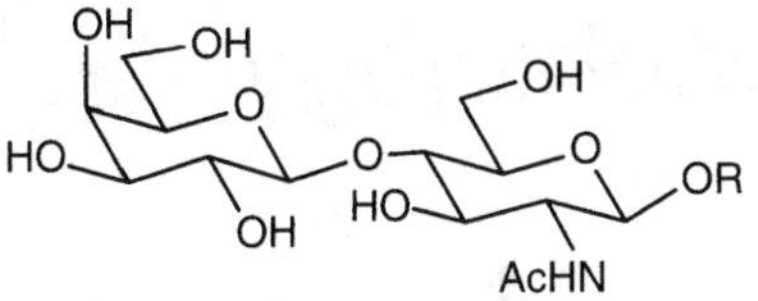

(b)

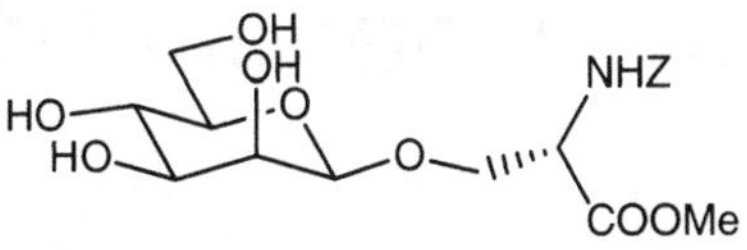

2. Through protein engineering of retaining β-galactosidases, new enzymes are formed that no longer possess their active site nucleophilic carboxylate but still have their active site acid/base carboxylate residue. By examining Fig. 8.10, explain by giving mechanisms the following observations:

    (a) These enzymes can no longer cleave *O*-linked disaccharides.
    (b) These enzymes catalyse the reaction of α-galactosyl fluorides with alcohol acceptors to give β-galactosides at higher pH.

3. *para*-Nitrophenylgalactoside **8.21** reacts with β-galactosidases to give a permanent glycosyl-enzyme adduct. Predict the structure of this adduct and explain why it is stable.

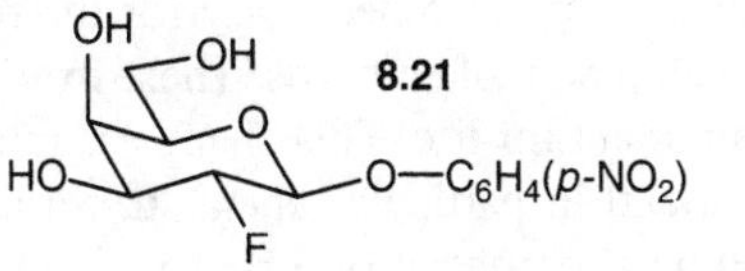

8.21

9 Chemical glycobiology

9.1 The importance of sugars in biology

Sugars are ubiquitous in Nature as one of the major classes of biomolecules. Although it has long been clear that sugars are both important foodstuffs and structurally important compounds, exciting developments in carbohydrate science over the past 20 years have started to reveal that they are involved in an enormous range of very precise and sophisticated processes. There are now numerous examples of sugars acting as signals for biological communication both within cells and between cells. For example, the altered structure of cell surface sugars has been implicated in tumour growth, and is involved in the spreading of cancerous cells around the body. The correct carbohydrate structures on cell surfaces are also necessary for the fertilisation of eggs by sperm and for the development of embryos. These examples clearly show that carbohydrates are important for life. This chapter is intended to illustrate that carbohydrate science, and in particular chemistry, truly has the potential to allow the understanding and control of a range of biological processes.

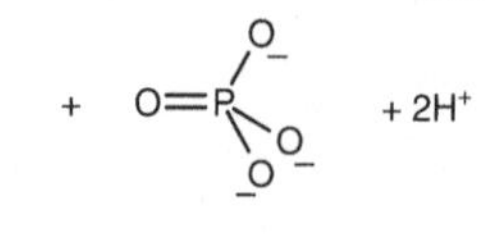

9.2 Metabolism: glucose as a source of energy

Adenosine triphosphate (ATP) **9.1** is the most important and rapidly available source of energy in biological systems. This energy is stored in the phosphodiester bond of ATP and is used to displace otherwise unfavourable biological equilibria and in doing so allows functions as varied as motion, active transport, biosynthesis, and signal amplification to be performed.

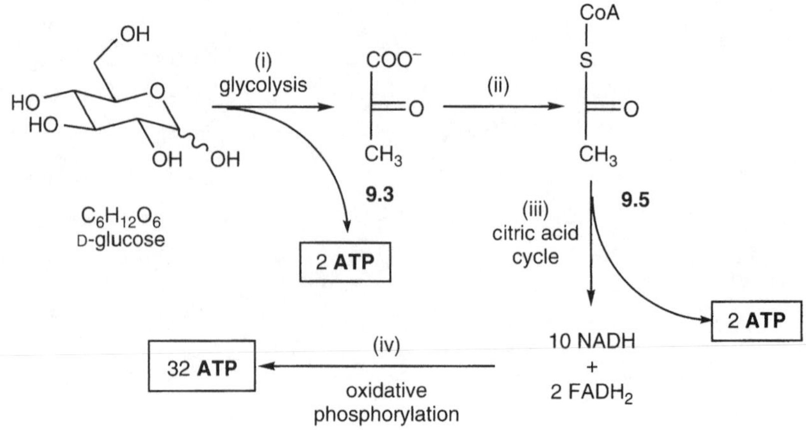

30.6 kJ mol^{-1} of free energy are released upon the hydrolysis of adenosine triphosphate (ATP) to adenosine diphosphate (ADP)

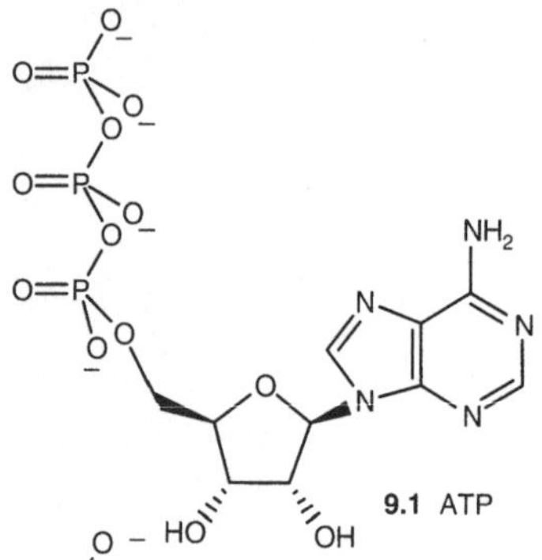

9.1 ATP

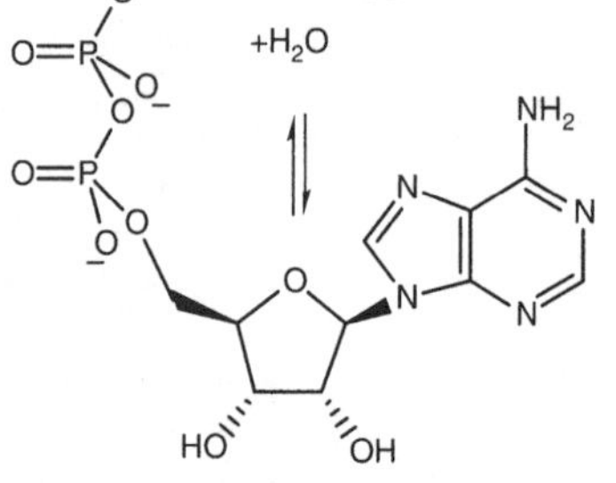

$+H_2O$

9.2 ADP

The metabolism of D-glucose is a highly efficient process. By comparing the free energy of combustion ($\Delta G° = -2881$ kJ mol^{-1}) with 36 ATP phosphodiester bond energies (each $\Delta G° = -30.6$ kJ mol^{-1}) that are the products of metabolism, we arrive at an efficiency of 38%.

$$\frac{-30.6 \times 36}{-2881} \times 100 = 38\%$$

$C_6H_{12}O_6$
D-glucose

(i) glycolysis

9.3

2 ATP

(ii)

9.5

(iii) citric acid cycle

2 ATP

10 NADH
+
2 FADH$_2$

(iv) oxidative phosphorylation

32 ATP

The major method for the regeneration of ATP **9.1** from ADP **9.2** is through the oxidation of foodstuffs. D-Glucose is the most important of these and the free energy that it contains is used to regenerate ATP in the four-stage process shown above.

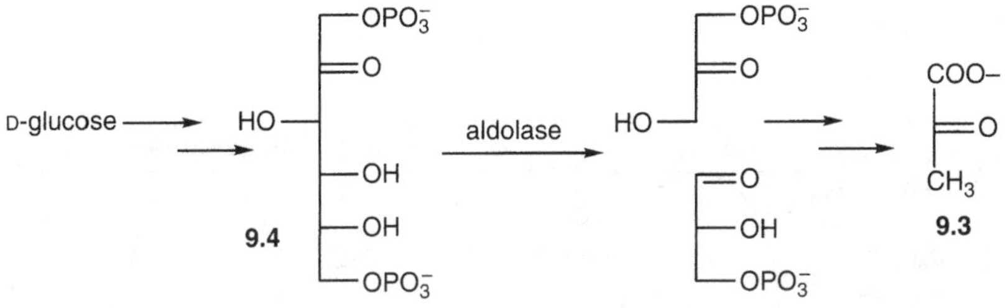

Fig. 9.1

Detailed knowledge of the structures of SCoA, NADH, and FADH$_2$ is not necessary to an understanding of the overall idea of glucose metabolism. However, it is useful to realise that NADH is formed from NAD$^+$ through reduction and thus acts as an electron carrier.

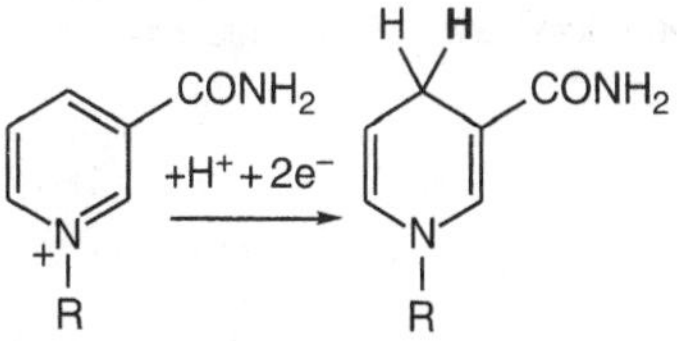

NADH can also be treated as the biological equivalent to H$^-$. The reduction of acetaldehyde to ethanol in fermentation is entirely analogous to the reduction of an aldehyde by a hydride-delivery reagent such as

$$CH_3CHO \xrightarrow[\text{NADH}]{\text{enzyme}} CH_3CH_2OH$$

cf

$$CH_3CHO \xrightarrow{\text{NaBH}_4} CH_3CH_2OH$$

sodium borohydride.
See Oxford Primer No. 20, John Mann, *Chemical Aspects of Biosynthesis* for further details.

Stage (i) is **glycolysis** (Fig. 9.1) which results in the splitting of one molecule of D-glucose into two of pyruvate **9.3**. The key stage in this ten-step sequence is a retroaldol reaction that is catalysed by an aldolase enzyme and results in the formation of two 3-carbon fragments from the 6-carbon sugar fructose-1 : 6-bisphosphate **9.4**. These fragments are then isomerised and dephosphorylated to give pyruvate **9.3**.

Stage (ii) is the conversion of pyruvate into an activated thioester of acetic acid known as acetyl-SCoA **9.5**. The SCoA portion of **9.5** is a good leaving group and so acetyl-SCoA effectively acts as Nature's equivalent to a synthetic chemist's acetyl chloride or acetic anhydride, and allows the entry of the acetyl group into the third stage.

Stage (iii) is the **Citric Acid Cycle** (CAC, also known as Krebs' Cycle or the Tricarboxylic Acid Cycle) which converts acetyl-SCoA into the reducing compounds **n**icotinamide **a**denine **d**inucleotide in its reduced form (NADH) and **f**lavin **a**denine **d**inucleotide in its reduced form (FADH$_2$). Because of their high reductive potential these act as effective carriers of electrons through to the final stage of glucose metabolism. The CAC achieves a remarkable feat, in that as well as using acetyl-SCoA to produce three NADH molecules and one FADH$_2$ molecule, it also utilises water as a source of reductive potential.

The final stage (iv) is **oxidative phosphorylation** (also known as the **electron transport chain**) which takes the 10 NADH and 2 FADH$_2$ molecules generated in stages (i)–(iii) and converts them into 32 molecules of ATP.

9.3 Fermentation

The process of metabolism described in the previous section requires a plentiful supply of oxygen. When this is absent due to lack of supply, for example, in muscles at times of rapid exertion, an alternative **anaerobic** process (to contrast it with the usual **aerobic** process) takes place. In yeast this alternative process has evolved to such a level that yeast may live even in the absence of oxygen. This is the basis of fermentation for the preparation of alcohol.

The term dehydrogenase (which really refers to an oxidation of a substrate by removal of hydrogen) is commonly used for enzymes which catalyse both oxidations and reductions. Whether the substrate is oxidised or reduced depends on the substrate and the cofactor used.

Notice that the process of converting pyruvate to ethanol regenerates a molecule of NAD$^+$ which is required for glycolysis. As a result of this recycling of NADH, yeast are able to continually produce alcohol from D-glucose.

The first stage of this process is glycolysis leading to the production of pyruvate by the same mechanism described in Section 9.2. Unlike in humans, in yeast the pyruvate produced is converted by an enzyme called a decarboxylase, which as its name implies splits pyruvate into CO_2 and acetaldehyde. The acetaldehyde is then reduced to ethanol by an enzyme called a dehydrogenase using the cofactor NADH.

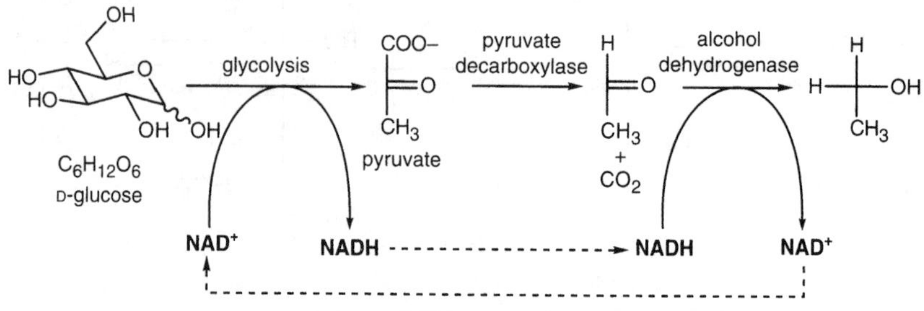

Fig. 9.2

It can be seen from Fig. 9.2 that the simple incubation of yeast with D-glucose ultimately produces CO_2 and alcohol; this is **fermentation**.

9.4   Structural building blocks: plants and lobsters

Nature uses sugars as building blocks in polymeric arrangements that provide a high level of structural strength. Starch **9.6** is a polymer of D-glucose in which each unit is linked to the next by a bond with α stereochemistry from the anomeric centre of one glucose to the oxygen atom of position 4 of the next. Starch is one of the main sources of D-glucose in our diet. Cellulose, **9.7**, is also a polymer of D-glucose in which the only difference from starch is that each unit is linked to the next by a bond with β rather than α stereochemistry at the anomeric centre. In plants, a coating of the polysaccharide cellulose protects the outer wall of cells. As a result of its structure, cellulose forms long linear chains that are many thousands of repeating units in length. These sugary strings assemble into larger well-ordered fibres through hydrogen bonding and van der Waals' forces. Cellulose is valuable to plants because of its chemical inertness and its complete insolubility in water. It is also very resistant to acid hydrolysis. In fact, cellulose is the most abundant organic compound in the plant kingdom and as such the most abundant carbohydrate on the planet. Mankind has harnessed the strength of cellulose in materials such as cotton and viscose rayon as well as esters of cellulose in plastics.

The polysaccharide chitin **9.8** is structurally similar to cellulose and is also a very strong and inert substance. It makes up the hard exoskeletons of shellfish and insects. Chitin is based on a repeating N-acetylglucosamine unit instead of a glucose repeating unit and so differs from cellulose only in the presence of an acetamido group at the 2 position.

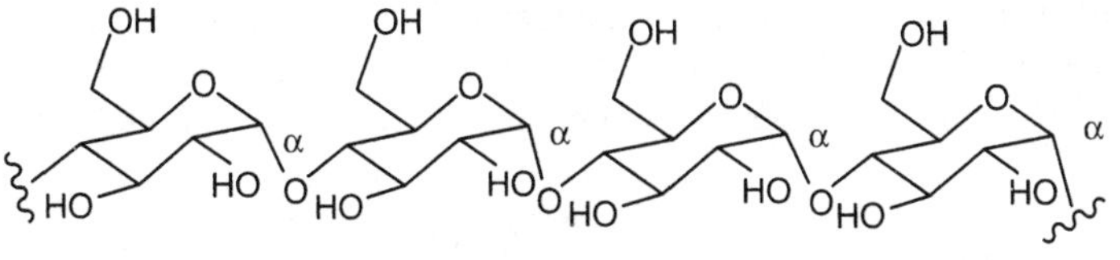

9.6 starch

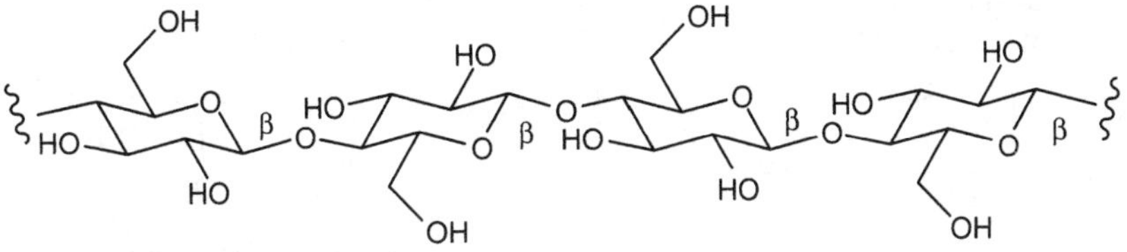

9.7 cellulose

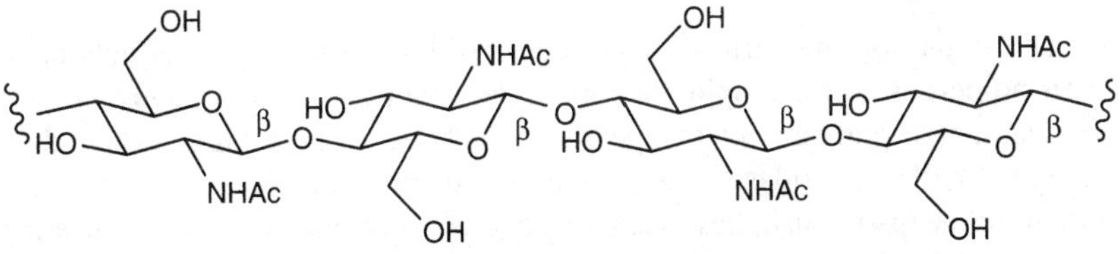

9.8 chitin

9.8

9.5 Glycocode

Carbohydrate structures are unrivalled in the density of information that they can convey. The nature of the linkages between two residues can be highly varied. For example, two pyranose molecules can be linked in five different ways via 1–1, 1–2, 1–3, 1–4 and 1–6 linkages. Compare this with the linear nature of proteins and nucleic acids. Add to this the additional variety caused by other variables such as the stereochemistry of the anomeric centre, ring size (e.g. pyranose or furanose), and modification of sugar hydroxyl groups (e.g. sulfation, phosphorylation, methylation, or acylation) and it can be quickly seen that this great variety of possible combinations gives the language of carbohydrates exquisite eloquence. This language has been christened **glycocode**—a term that well represents the high level of complex information that carbohydrate structures can convey.

A comparison of the permutations of hexamer formation illustrates this point well. Whereas, DNA (with a basis set of 4) and amino acids (with a basis set of 20) may construct a biological language for information transfer of 4096 and 6.4×10^7 'words', respectively, carbohydrates have access to greater than 1.05×10^{12} variations. With all these permutations there are far too many possible oligosaccharides for one to consider making them all. Instead, the design of new carbohydrate-containing structures succeeds by identifying the functions of existing structures.

The multiple functional groups of carbohydrates allow them to be assembled in a large number of ways.

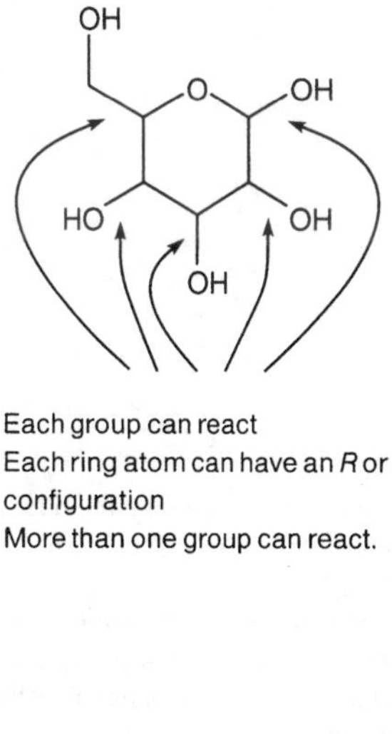

- Each group can react
- Each ring atom can have an *R* or *S* configuration
- More than one group can react.

Even if we were able to make a different hexasaccharide every day, the world would still end before we made all the possibilities.

9.6 Glycoproteins

Most of the proteins synthesised in mammalian cells have carbohydrates attached to them. In these **glycoproteins** carbohydrates are attached covalently to either the amide nitrogen atom in the side chain of the amino acid asparagine, (*N*-linked glycoproteins **9.9**) or to the oxygen atom in the side chain of the amino acids serine or threonine (*O*-linked glycoproteins **9.10**).

9.9

9.10

R = H	for serine
= CH₃	for threonine

The sugars on the surface of proteins can have a variety of effects upon their properties. For example, they increase solubility, make proteins more or less antigenic and protect the peptide backbone against protein-degrading enzymes called peptidases. A fascinating function of glycoproteins in the blood of deep-sea fish has been discovered. Proteins with the repeating structure **9.11** allow these fish to survive at temperatures as low as $-2\,°C$ due to an antifreeze effect that is thought to arise from the inhibition of ice crystal growth by the hydroxyl groups of the sugars on these proteins.

9.11

9.12

An example of an oligosaccharide present on *N*-linked glycoproteins. Although this structure does not represent the true *conformation* of this oligosaccharide, it does give an idea of the branched, 'hand-like' nature of the sugar portion of glycoproteins. It is easy to see why the docking of these sugars with binding proteins is known as a 'handshaking' process.

Branched sugars on glycoproteins are the manifestation of the code we discussed in the previous section. *N*-linked and *O*-linked glycoproteins contain different core sugar structures but complex and branched biologically active oligosaccharides are often found on outer chains attached to these cores, for example, **9.12**. These act as sources of information to sugar binding proteins such as lectins (see Section 9.8). For example, in certain blood glycoproteins, for example, **9.12**, the presence of a sialic acid sugar residue at the tip of these branches is normal. However, as these proteins 'age' degradation leads to the loss of this sialic acid cap, which exposes the D-galactose residue underneath. This D-galactose residue is recognised by a protein on the surface of the liver called the asialoglycoprotein receptor (*asialo* = without sialic acid). This binds and absorbs the 'old' blood glycoprotein and the cell to which it is attached into the liver for degradation. Thus, the information or code transmitted by the sugars in this example is 'new' (= with sialic acid) or 'old' (= without sialic acid).

Unlike the biosynthesis of proteins and nucleic acids there appears to be no associated mechanism for proof reading and correcting differently glycosylated proteins—the result is that they are formed as mixtures. Therefore, glycoproteins occur naturally in a number of forms that have the same peptide backbone, but with different sugars and different sites of glycosylation. These different forms are called **glycoforms**. Each of these glycoforms has a different property. It has been suggested that these naturally occurring mixtures of glycoforms provide a spectrum of activities that can be carefully altered by the body in one direction or another as a means of fine-tuning protein properties.

9.7 Glycolipids

Glycolipids are compounds with both a polar (sugar) and a nonpolar (lipid) portion. Like glycoproteins, they are commonly found components of the plasma membrane of all vertebrate cells. They provide organisms with another method for the attachment of sugars to the surface of cells. Instead of being covalently linked to cell surface proteins, in the way that glycoproteins sugars are attached, the hydrophobic lipid part of glycolipids buries itself in the outer lipid layer of the cell with the hydrophilic sugar portion pointing outwards from the cell (Fig. 9.3). Multiple copies of glycolipids are found on surfaces of cells in 'rafts'. These 'rafts' are formed through lateral movement of glycolipids over the cell surface to form clusters. This 'sliding together' allows many branched sugars to be displayed. This is similar to the way that glycoproteins can also present many oligosaccharide structures, by using several amino acid residues as glycosylation sites. An example of one type of glycolipid is the cerebroside **9.13**, so-called because this type of glycolipid is found in high quantities in the brain, which contains the sugar β-D-glucose linked to a fatty nonpolar lipid called a ceramide.

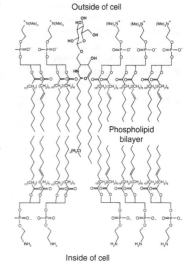

Fig. 9.3

9.13

$n = 6, 8, 10, ...$

More than 400 types of glycolipids possessing different sugar structures have been reported, although only 7 monosaccharides are commonly found in vertebrate glycolipids. The significance of the lipid part is still not understood well. It is thought that subtle variations in the structure of the lipid might influence the position and functions of glycolipids on the cell's plasma membrane, possibly by interaction with cholesterol, phospholipids, and the hydrophobic portions of receptor proteins.

The balance of the biosynthesis and degradation of glycolipids needs to be carefully regulated in the cell. If a glycosylhydrolase that catalyses glycolipid degradation is lacking due to a genetic deficiency, glycolipids accumulate in the lysosomes of the cell and cause serious disease. Certain drugs, such as the enzyme inhibitor NB-DNJ (see Section 9.13) which blocks the synthesis of glycolipids, offer the potential for treating some of these so-called glycolipid storage diseases, such as Gaucher disease and Tay–Sachs disease. The biological functions of glycolipids, as well as the mechanism that regulates glycolipid metabolism, are still not well understood but it is thought they too are another source of sugar communication molecules. Carbohydrate chemistry has allowed the synthesis of a number of naturally occurring glycolipids as valuable biological probes using the glycosylation techniques described in Chapters 7 and 8.

9.8   Lectins

The term lectin, which owes its origin to the Latin word *legere* meaning specific, was first used by Boyd in 1954 to describe proteins that show a potent and highly specific ability to bind glycosylated structures. The term has subsequently been redefined to describe carbohydrate-binding proteins that are neither enzymes nor antibodies (although some similarities in modes of binding and some intermediate cases have meant that this distinction is starting to be questioned).

The decipherers of **glycocode** (see Section 9.5) are typically sugar-binding proteins called lectins, which despite their very shallow binding sites, show a remarkable specificity in their recognition of highly-branched, complex carbohydrates. The binding that happens in the **carbohydrate recognition domain** (CRD) of lectins is largely due to hydrogen bonding between backbone and side chain carbonyl groups in the protein, and the hydroxyl groups of the sugar ligands (Fig. 9.4). Protein-bound calcium ions can also chelate vicinal hydroxyl carbohydrate groups. Van der Waals' interactions between hydrophobic lectin protein residue side chains and hydrophobic 'patches' on carbohydrates also increase binding affinity.

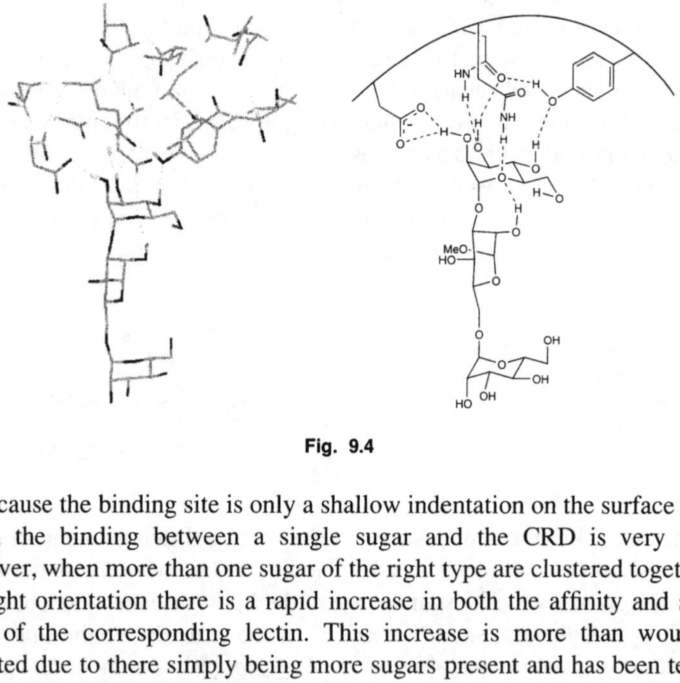

Fig. 9.4

Because the binding site is only a shallow indentation on the surface of the lectin, the binding between a single sugar and the CRD is very weak. However, when more than one sugar of the right type are clustered together in the right orientation there is a rapid increase in both the affinity and specificity of the corresponding lectin. This increase is more than would be expected due to there simply being more sugars present and has been termed the **cluster** or **multivalent effect**. There is usually only one way that a particular branched sugar structure can efficiently fit into the CRD. This is the source of the lectin's specificity. In a sense, branched oligosaccharides of glycoproteins or glycolipids act like sugary 'hands' with each sugar being a 'finger tip'. Many lectins have several CRDs which act as the 'finger holes' in a lectin 'bowling ball' which can only easily be gripped when all fingers are in place, for example, the mannose-binding snowdrop lectin has three CRDs on its surface. Alternatively, a similar effect can be achieved on a cell surface by having many copies of lectin proteins each with just a single CRD, for example, the asialoglycoprotein receptor in the liver, which forms hexamers.

9.9   Carbohydrates in inflammation

Damage in tissues surrounding a blood vessel causes the release of signalling molecules that start something called the inflammatory response. Lectins called selectins are rapidly formed on the inner surface of blood vessels (endothelia, hence **E**-selectin) and on **p**latelets (**P**-selectin), which strongly bind complex carbohydrates in a specific manner. This binding causes white blood cells (leucocytes) to stick to the walls of blood vessels. A two-way binding process is completed by **L**-selectins on the leucocytes. In the flow of blood, the leucocytes roll towards the site of damage, slowing down as lectin binding increases, where they pass through to the surrounding tissue (Fig. 9.5).

sialyl Lewis-x
sLe^x

Rolling from
selectin to selectin
with blood flow

Extrovasation
once stopped

selectin CRD

Fig. 9.5

Whilst this is a well-controlled process in healthy individuals, in excess it is a cause of septic shock, arthritis, asthma, and heart disease. The precise identity of the sugar that selectins bind to is not known. However, the tetrasaccharide sialyl Lewis-x (sLe^x) **9.14** is specifically bound by all types and serves as a useful benchmark. By making carbohydrates that block the binding of sLe^x to selectins, drugs can be designed to prevent these diseases. For example, **9.15** shows an 800-fold more potent inhibition of P-selectin than sLe^x **9.14**.

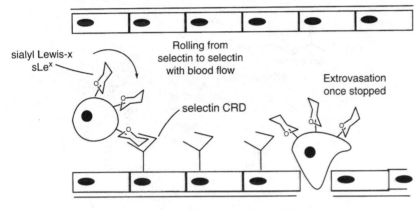

9.14

sialyl Lewis-x (sLe^x)

9.15

9.10   Carbohydrate vaccines

The use of complex oligosaccharides for the induction of antibodies has a rich history dating back to 1930 when a vaccine against pneumonia was made by attaching oligosaccharides to proteins. Unfortunately, this work was limited by the often-minute amounts of oligosaccharides available from natural sources. It was not until the 1970s when the synthesis of oligosaccharides was simplified by some of the methods described in Chapter 7, that things became easier.

Things have now advanced to the point that we can design potential anti-cancer vaccines. The sugar sialyl-Tn (or sTn) **9.16** appears in higher than normal levels on the surface of breast cancer cells. Attaching **9.16** to a protein creates a synthetic vaccine that produces high levels of antibodies against cancerous cells. Excitingly, this anti-tumour vaccine causes higher survival rates of cancer patients in clinical trials.

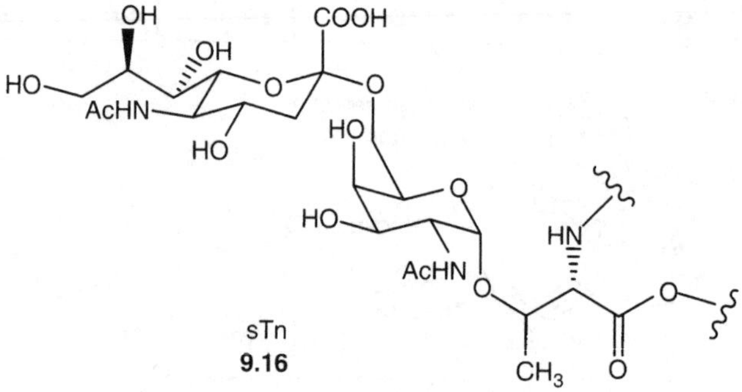

sTn
9.16

9.11   Sugars and infection by microorganisms

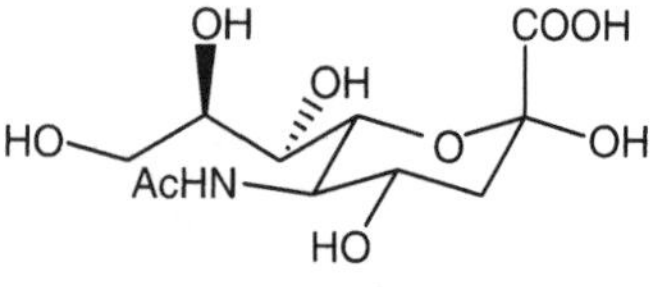

sialic acid
9.17

A disadvantage of the formation of carbohydrates at cell surfaces is that sugars on cells lining the trachea, stomach, and intestines can help micro-organisms and viruses to invade the body. Thus, carbohydrates are sometimes a gateway for foreign organisms. For example, infections by the influenza virus, the bacteria that causes stomach ulcers (*Helicobacter pylori*), and the parasite protozoan that causes Chagas' disease (*Trypanosoma cruzi*) all rely on the binding between lectins on cell surfaces and the sugar sialic acid **9.17**. Interestingly, human milk is full of sialic-acid-containing compounds. It has been suggested that they protect newborn infants by acting as decoys for these foreign organisms in the gut, which bind to these compounds instead and are thus passed out of the body. Carbohydrate chemists have learnt this lesson from Nature and have used this idea to design a number of drugs containing sialic acid which work in a similar way.

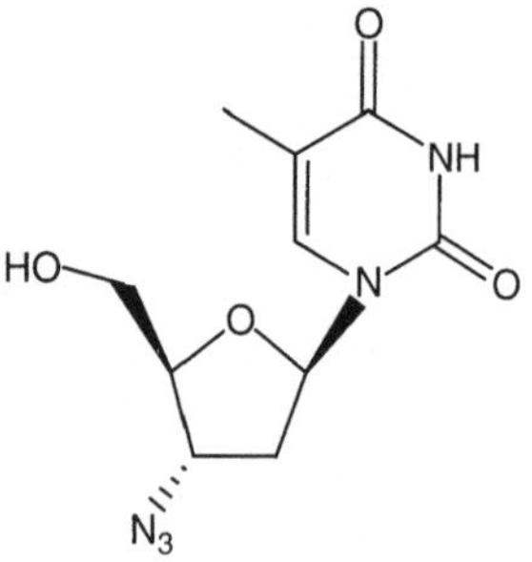

9.18 AZT

9.12   AZT

Because they are an important class of biological molecules in their own right, it's easy to forget that ribonucleic acids (RNA) and 2′-deoxy-ribonucleic acids (DNA) are also carbohydrate compounds containing the pentose D-ribose. One synthesis that relies heavily on carbohydrate chemistry

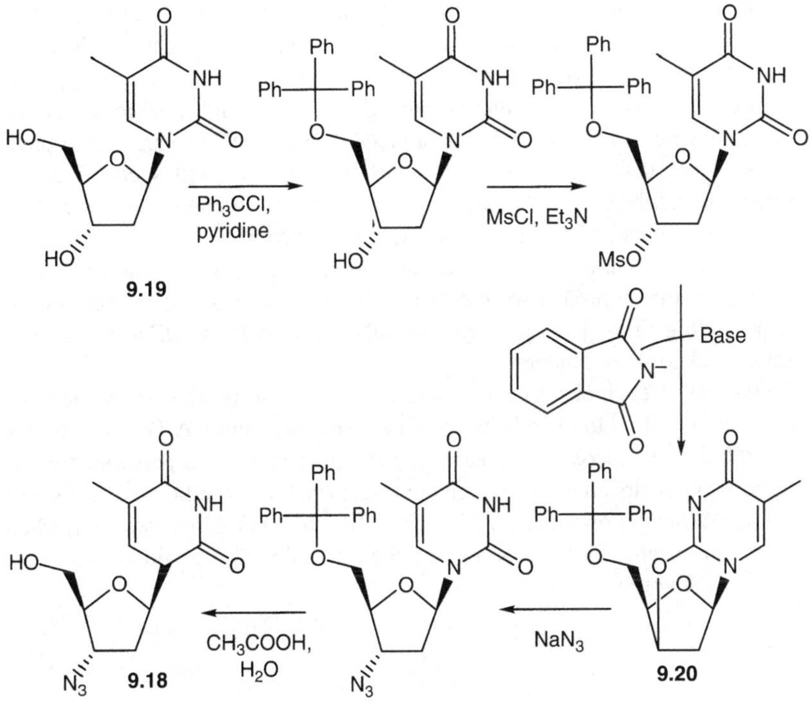

Fig. 9.6

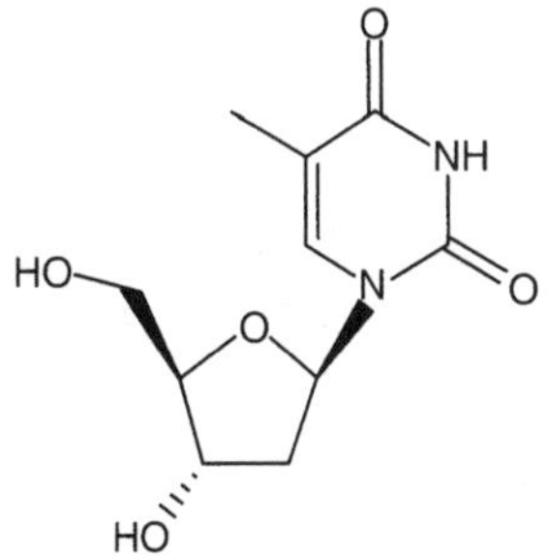

Notice how the configuration of the carbon atom that we introduce the azide group to is the same as that in the starting material **9.19**. This is a result of two inversions of configuration, the first by the oxygen atom of the thymine group, which displaces the mesylate at C-3′ to form **9.20**, and the second by azide ion. Double inversion = overall retention.

is that of the very important anti-HIV drug, 3′-azido-3′-deoxythymidine (AZT) **9.18** which acts by resembling the DNA molecule deoxythymidine **9.19**. The synthesis of AZT **9.18** from **9.19** is shown in Fig. 9.6 and uses some of the methods for the protection and reaction of hydroxyl groups that were described in Chapter 4.

AZT **9.18** is a **reverse transcriptase** inhibitor. Viruses like HIV are **retroviruses**, so-called because they contain RNA rather than DNA and construct their own DNA using an enzyme carried within the viral particle called *reverse transcriptase*. This enzyme readily accepts a wide range of nucleotide triphosphate analogues. In the body, AZT **9.18** is converted to its triphosphate analogue and because it resembles the natural 2′-deoxyribonucleic acid deoxythymidine **9.19** it is readily incorporated by reverse transcriptase into the chains of newly formed HIV DNA. DNA is synthesised from its 5′ to its 3′ end by the addition of nucleotides step-by-step to the 3′-hydroxyl group in the nucleotide at the end of the growing DNA chain. Because AZT does not contain a 3′-hydroxyl group and has an azide group at this position instead, the chain cannot be elongated and correct DNA synthesis fails. The virus cannot produce its DNA and so its reproductive cycle is halted. AZT is just one of a whole host of drugs that act by resembling DNA or RNA molecules which rely on sugar chemistry for their synthesis.

9.13 Glycosidase and glycosyltransferase inhibitors

In the last chapter we saw that the two types of Carbohydrate Processing Enzymes—glycosidases and glycosyltransferases—are powerful catalysts for the synthesis of sugars. They are the tools that Nature uses to construct all of the carbohydrate structures that we have discussed in this chapter. With the idea of being able to control the biosynthesis of these structures, carbohydrate scientists have designed a number of inhibitors of such enzymes. As we have discovered, it is essential for branched sugars on cell surfaces to have the correct structure and shape in order for them to bind successfully to lectins. By inhibiting particular enzymes in the biosynthetic pathways that lead to these branched sugars on, for example, glycoproteins, the structures of these sugars are subtly but crucially altered. This causes the binding to fail.

Through this approach we can begin to treat diseases that rely on lectin-sugar binding. Two examples are NB-DNJ **9.21** to treat AIDS and swainsonine **9.23** to treat cancer.

NB-DNJ **9.21** looks like D-glucose **9.22** and it is able to reduce the infection of white blood cells by the HIV virus that causes AIDS. It does this by inhibiting a glucosidase that is one of the enzymes responsible for the synthesis of sugars attached to a glycoprotein (gp120) found on the surface of HIV. Gp120 is important in the binding of HIV to white blood cells and when its structure is altered, for example, through the inhibition of glucosidase, this binding is severely reduced.

If we compare the structure of the D-glucosyl cation **9.25** intermediate of glucosidase with NB-DNJ **9.21** we can see how they resemble each other.

9.25   at body pH nitrogen atom is protonated and so looks like O⁺

9.21   around the ring stereochemistry of **9.25** is the same as that of **9.21**

Note that if the stereochemistry of *any* of the hydroxyl groups in **9.21** is altered then the glucosidase is no longer inhibited. For example, **9.26** does not inhibit glucosidase at all.

9.26

Therefore carbohydrate drugs such as **9.21** are highly specific and consequently potentially free of certain side effects.

9.21   **9.22**   **9.23**   **9.24**

Strictly, NB-DNJ **9.21** inhibits glucosidases by mimicking the glucosyl cation intermediate **9.25** (or the transition state on the way to it) that is formed during the mechanism of action of glucosidases (see Chapter 8 for more details). This is a species that is tightly bound by the enzyme and so NB-DNJ is similarly tightly bound. In fact, NB-DNJ is so tightly bound that the active site of the enzyme is effectively blocked and unable to function. In the same way swainsonine **9.23** inhibits mannosidase because it resembles D-mannose **9.24**.

The design of glycosyltransferase inhibitors in a similar way has the potential to allow the synthesis of drugs that are even more specific and in order to develop treatments for a range of diseases this is just one of the exciting new goals for carbohydrate chemists.

9.14   Summary

At the end of this chapter you should be able:
- to explain the importance of D-glucose as a source of metabolic energy;
- to give an overview of the processes of D-glucose oxidation and how it leads to the formation of ATP;

- to explain fermentation as an alternative metabolic process found in yeast;
- to recall examples of structurally important polysaccharides;
- to explain with examples the concept of 'glycocode';
- to describe structures and functions of glycoproteins and glycolipids;
- to explain the mechanism by which lectins are able to act as highly specific carbohydrate-binding proteins;
- to describe the key biological roles that carbohydrates play in inflammation and infection by microorganisms;
- to recall examples of the use of carbohydrates and their mimics in the design of useful medical treatments such as vaccines, anti-HIV, and anticancer drugs;
- to summarise and give examples of the wide and varied roles of carbohydrate containing structures in biology.

9.15 Questions

1. Write brief notes on how carbohydrates act as communication molecules.
2. Why should chemists be interested in the biological functions of carbohydrates? Shouldn't they instead concentrate on making all the unsynthesised carbohydrate structures?
3. By drawing on features that you know to be successful in designing glucosidase inhibitors, design a potential sialidase (an enzyme that cleaves glycosidic bonds to sialic acid residues) inhibitor.
4. NB-DNJ **9.21** as well as being a glucosidase inhibitor also inhibits the biosynthesis of glycolipids. Can you suggest how it might do this on a molecular level?

Appendix: Drawing sugars

We shall use the example of D-glucose drawn in its pyranose form but the principle is the same for any sugar of any size, including furanoses.

Stage 1: Taking the Fischer projection and converting it into a 'pig trough'

Make sure you are happy with what a Fischer projection represents. It often helps to build a model the first time you try this. The Fischer projection is a representation of what you see from above. The carbon–carbon bonds above and below those in the chain point into the page and the bonds to the left and right stick out from the page.

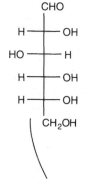

You can also draw a Fischer projection with the H atoms shown, like this. They are omitted in the Fischer projection on the right but added in the 'pig trough'.

To see what this looks like, in your mind's eye imagine the carbon backbone of the Fischer projection as the base of a 'pig trough' with the C-1 aldehyde (–CHO) and C-6 primary alcohol (–CH₂OH) groups forming the ends of the 'trough' directed away from you and the C–H and C–OH bonds on C-2, -3, -4, -5 making up the sides of the 'trough' sticking towards you. Viewed from a side angle, the 'trough' would look like that on the right.

Stage 2: Converting the 'pig trough' into a zigzag representation

In Chapter 2, we highlighted the importance of knowing how to convert Fischer representations into zigzag representations. Zigzag formulae place all of the carbon atoms C-1, -2, -3, -4, -5, -6 in the plane of the paper. This is, therefore, the first step of converting the 'trough' to the zigzag. This is best done by imagining the trough viewed directly from the side, that is, at 90° rather than at the side angled view we reached in Stage 1.

Once done, we can see how the C–OH and C–H bonds stick in and out of the plane of the page and we have represented this in the usual dotted and wedged fashion.

A 'pig trough' viewed
from the side angle

A 'pig trough' viewed
from the side

The final part is to convert the C-1, -2, -3, -4, -5, -6 carbon chain from a straight form to the zigzag we require. To do this imagine holding the C1–C2 bond steady and rotate the C2–C3 bond by 180°. As you can see, this introduces the first of our zigs. Next rotate the C3–C4 by 180° then the C4–C5 and so on.

A 'pig trough' viewed
from the side

Stage 3: Converting the 'pig trough' into a ring structure

The 'trough' is also a good starting point for working out how to draw the cyclised or ring structures of sugars.

rotate C4–C5
until OH is in the plane of the paper

A 'pig trough' viewed
from the side

A 'linked trough'
viewed from the side

Conversion to a pyranose structure requires the linking of the OH group at position 5 to the anomeric C-1 carbon. To do this we need to rotate OH-5 to a position in the plane of the paper so that it's in a good position to link to C-1.

This can be accomplished by rotation around the C4–C5 carbon–carbon bond. The poised OH-5 can be linked to C-1 to form a lactol (hemiacetal) in a 'linked trough' structure.

A 'linked trough'
viewed from the side

Topple the 'linked trough' towards you

α
Haworth projection

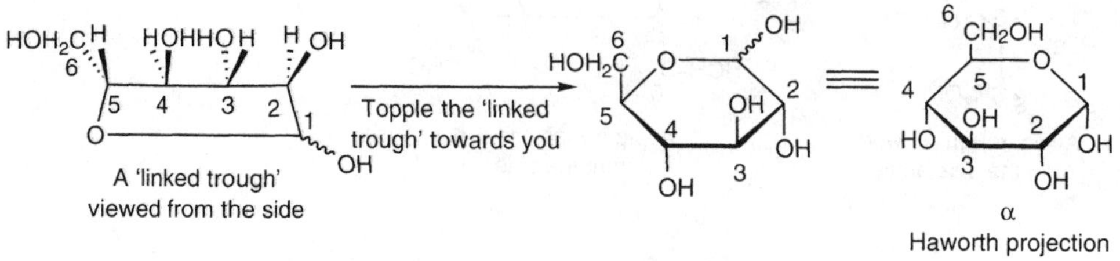

We have arbitrarily chosen to show the α-anomeric configuration here, by drawing the OH group at C-1 down.

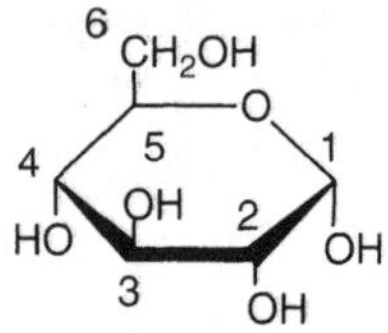

α

Alternatively, we could have chosen to show the β configuration.

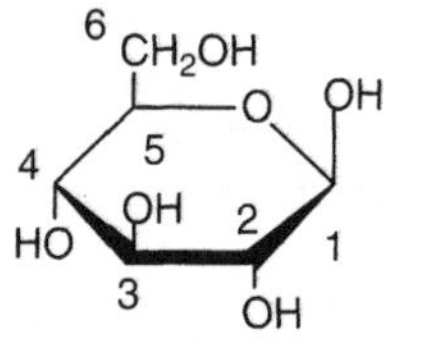

β

Don't draw chairs like this:

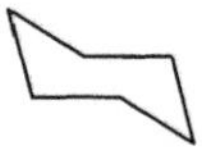

Not only does it look terrible but it also makes the substituent bonds harder to draw and axial bonds harder to distinguish from equatorial ones.

Finally, this 'linked trough' structure can be converted to a Haworth projection simply by 'toppling' this structure over towards you.

Stage 4: Drawing the chair form

Finally, in order to draw the chair in a more realistic chair conformation, we need to pucker the flat ring of the Haworth projection by raising the C-4 atom and lowering the C-1 to form a $^4C_1$ conformation. Note that for glucose all the bonds are equatorial except the anomeric C-1 bond, which is axial for α and equatorial for β.

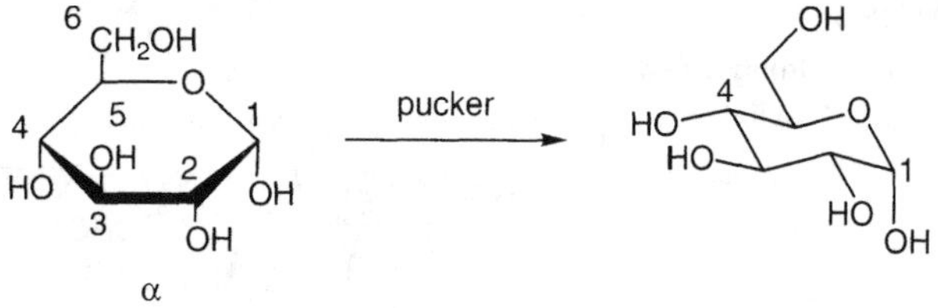

α

pucker

We could also have puckered the ring in the opposite way (by lowering the C-4 atom and raising the C-1 to form a $^1C_4$ conformation instead) but in this conformation too many bonds are axial and therefore 1,3-diaxial interactions make it energetically less stable. As a result the conformational equilibrium lies on the side of the form we first drew.

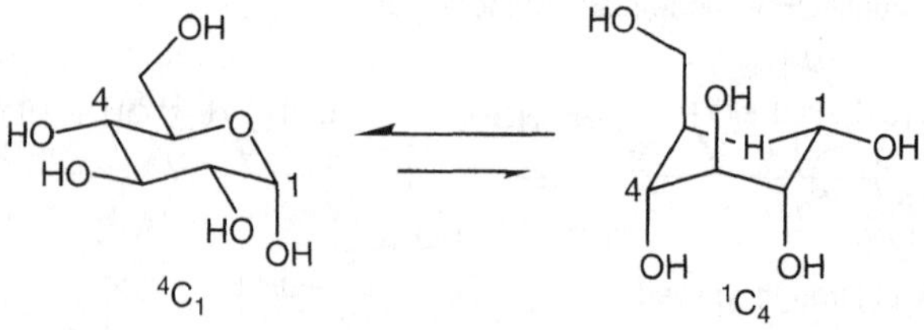

$^4C_1$

$^1C_4$

For 5-membered (furanose) rings the same 'linked trough' approach can be used except that in Stage 3 the C3–C4 carbon–carbon bond is rotated to place OH-4 in a down position suitable for linking to C-1.

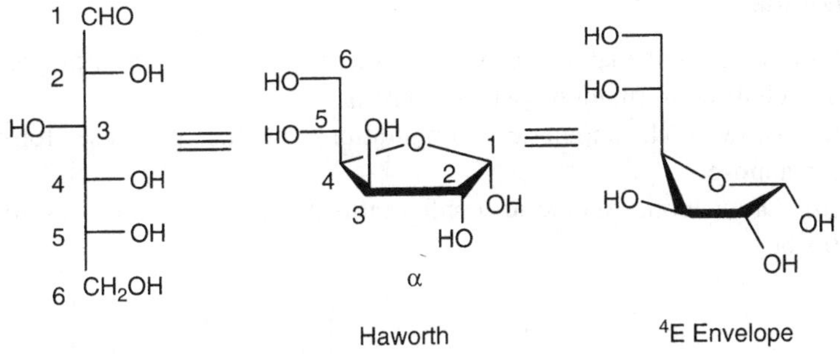

Haworth   ⁴E Envelope

In each drawing below the thickened bonds are parallel—use this to help you draw the conformation and substituent bonds of a chair correctly.

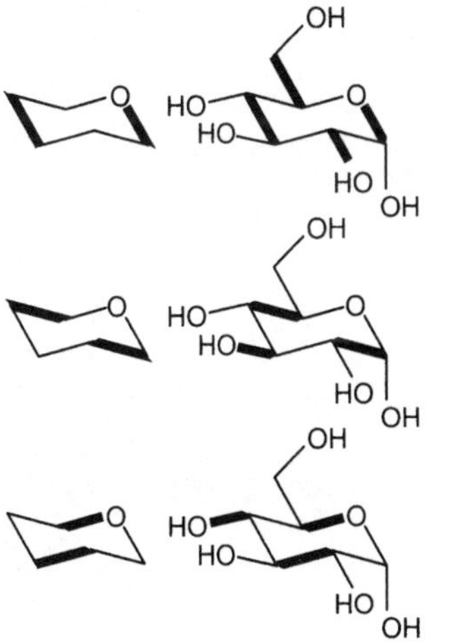

We tend to leave furanose forms in the 'unpuckered' Haworth projection as, unlike 6-membered rings, things are easier to manipulate in this form than their more realistic envelope conformation.

The LURD Trick

This handy mnemonic can be used to check if structures are drawn correctly:

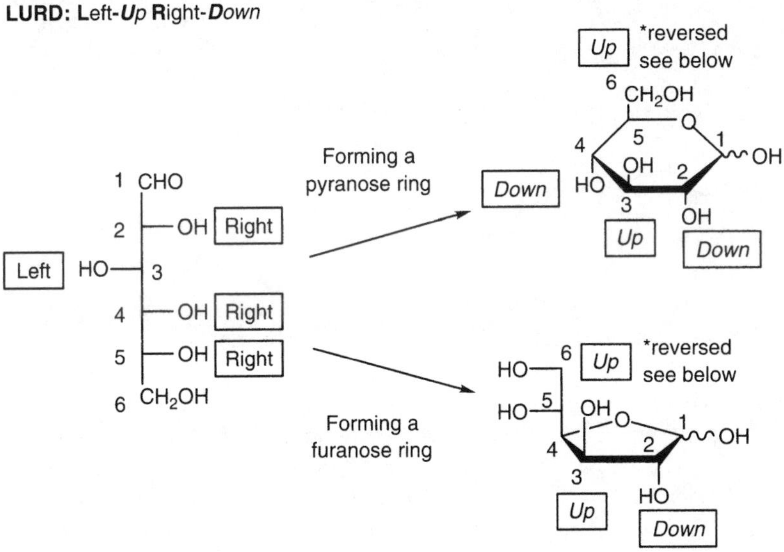

LURD: Left-*Up* Right-*Down*

* LURD rule is reversed at the carbon atom that forms the ring

Note that the LURD rule is reversed at the carbon atom that forms the ring. That is, what would have been a *Down* OH using the LURD trick in both of the examples shown here is actually an *Up* side chain instead.

For D-glucose, the Fischer projection has hydroxyl groups directed **Right–Left–Right–Right**. Therefore, using the **LURD** mnemonic, we would expect **Down–Up–Down–Down**. However, the rule is reversed at the atom that forms the ring. This gives us a final arrangement of **Down–Up–Down–Up** for the *substituents* in D-glucopyranose and **Down–Up–Up** for the *substituents* in D-glucofuranose.

Questions

1. Satisfy yourself that you can draw other sugars in Haworth projections and chair conformations, where relevant.

   (a) Draw both anomeric configurations of the pyranose form of L-mannose.

   (b) Draw both anomeric configurations of the furanose form of D-ribose.

(a) L-mannose   (b) D-ribose

Answers to problems

Chapter 2

2.1. (b) α- (64.9%) and β- (34.2%) D-mannopyranose and α- (0.6%) and β- (0.3%) D-mannofuranose.

2.2. (c) $^4C_1$: α has 1, β 0; $^1C_4$ α has 4, β has 5; (d) D-altrose, talose & gulose have 3 or 2 ax & 3 or 2 eq.—in fact α-D-altrose is the closest with a difference of 0.84 kJ mol^{-1} between $^4C_1$ and $^1C_4$.

Chapter 3

3.1. Methanolysis of $CH_3COCl/SOCl_2$ forms the corresponding ester and HCl.

3.2. The axial orientation is favoured by the anomeric effect, but disfavoured by steric effects. In CCl_4 the anomeric oxygen is not H-bonded to the solvent, so is small, and the anomeric effect predominates. In water, its effective size is increased by H-bonding and, therefore, more of the equatorial form is present at equilibrium.

3.3. Only the anomeric methyl group is hydrolysed via an S_N1 process to yield tetra-O-methylglucopyranose as a mixture of α and β anomers.

3.4. Equilibration occurs via the mutarotation mechanism via the open chain aldehyde. Re-closure of the aldehyde happens faster than exchange of the aldehyde oxygen with the solvent via hydration/dehydration.

3.5. Fischer glycosidation to give methyl α-glucopyranoside because of the anomeric effect.

3.6. **I** forms a glycosyl cation, **II** is the kinetic product in which the anomeric effect results in an axial O, **III** is the thermodynamic product *trans* decalin-like structure (no 1,3-diaxial interaction).

Chapter 4

4.1. (i) Make the methyl glycoside by Fischer glycosidation with methanol and acid, then benzylate all free hydroxyls with benzyl bromide and sodium hydride, then hydrolyse the methyl glycoside with strong aqueous acid, (ii) Make the methyl glycoside, selectively iodinate primary hydroxyl group, reduce, then hydrolyse the methyl glycoside with strong acid. Alternatively protect the primary hydroxyl, for example, trityl ether, protect others (e.g. acetate), remove trityl, mesylate, displace with iodide, reduce and then remove all protecting groups (base then strong acid).

4.2. Diaxial nucleophilic epoxide opening. **A** is attacked by nucleophiles at C-3, **B** at C-2.

4.3.   Selectively tritylate primary hydroxyl, mesylate the other two. Displace both mesylates with iodide with inversion to give the diaxial iodide, which reductively eliminates to give alkene. Finally base hydrolyses the benzoyl ester.

Chapter 5

5.1.   (i) Make the peracetyl bromide, glycosylate with methanol then remove acetates, (ii) Use degradation, for example, via the *bis*-sulphone, (iii) Make the 1,2-orthoester from the peracetyl bromide, remove acetates, replace with benzyls, and hydrolyse the orthoester in aqueous acid, (iv) Use the Kiliani ascension and lactonise with mild acid.

5.2.   Make the anomeric bromide and reductively eliminate to make triacetyl glycal. Remove the acetates and replace with benzyls. The cyclic iodonium ion on the top (β) face is diaxially opened at the anomeric centre by methanol to give the 2-iodo methyl glycoside with α-manno stereochemistry.

5.3.   Glucitol using $NaBH_4$, 1-deoxysugar using Bu_3SnH reduction on per-acetylbromide then remove acetates.

Chapter 6

6.1.   5-Ring *cis* acetonide formation, and simultaneous Fischer glycosidation with methanol released from reagent.

6.2.   First, formation of diacetone galactose, then methylation of the 6-hydroxyl followed by hydrolysis of both acetonides, and formation of the methyl α-glucopyranoside by Fischer glycosidation with methanol.

6.3.   (i) Form diacetone glucose with acetone/acid, then benzylate the free 3 hydroxyl. Treat with strong aqueous acid to hydrolyse both acetonides and get back to the pyranose form, (ii) Form diacetone glucose with acetone/acid, then oxidise (PCC) the free 3 hydroxyl, then reduce ($NaBH_4$) to effectively epimerise. Finally treat with strong aqueous acid to hydrolyse both acetonides and get back to the pyranose form, (iii) Form diacetone glucose with acetone/acid, then selectively hydrolyse the primary acetonide with aqueous acetic acid. Then periodate cleave between C-5 and C-6 to give the C-5 aldehyde. Reduce and then treat with strong aqueous acid to hydrolyse the remaining acetonide and get back to the pyranose form of the shortened sugar.

6.4.   Make the 4,6-benzylidene, then benzylate the free 3-hydroxyl. Hydrolyse the benzylidene in mild acid, then selectively silylate the primary hydroxyl with the bulky silyl group.

6.5.   Only the 3 and 4 equatorial hydroxyl groups react to give a 6-ring cyclic diacetal (cyclohexane diacetal, CDA) as for butane diacetal formation, which is stabilised by multiple anomeric effects.

Chapter 7

7.1.   (i) Gives α-mannose linked to the 6-hydroxyl of diacetone galactose due to neighbouring group participation, (ii) Gives β-glucose linked to the 6-hydroxyl of diacetone galactose, again stereocontrol by neighbouring group participation.

7.2. Make methyl α-mannopyranoside with methanol and acid, then selectively protect the 6-hydroxyl (e.g. silyl), then make 2,3-acetonide leaving OH-4 free—this is the acceptor. Make peracetylmannose bromide—this is the donor. Glycosylate, remove all protecting groups by treatment with mild acid then methoxide.

7.3. **7.16** can react with acceptors to give unwanted orthoester; **7.17** has a sterically crowded orthoester carbon and so does not.

Chapter 8

8.1. (a) β(1,4)galactosyltransferase with Gal-UDP or retaining β-galacto-sidase with β-galactoside donor and excess *N*-acetylglucosamine acceptor, (b) β-mannosyltransferase or β-mannosidase with β-manno-side donor and Z-protected L-serine acceptor.

8.2. (a) No nucleophile means that even though the leaving group is pro-tonated, the α-glycosyl-enzyme intermediate **8.14** is not formed, (b) The α-glycosyl fluoride is sufficiently activated to act as a mimic of the α-glycosyl-enzyme intermediate and so the acceptor attacks from the β-face aided by the enzyme carboxylate acting as a general base.

8.3. The adduct is a 2-fluoroglycosyl-enzyme ester intermediate. The fluoride group withdraws electrons from the carbohydrate ring and so makes it less reactive. The additional use of the *para*-nitrophenyl leaving group creates an activated glycoside that is sufficiently active to react with the nucleophilic carboxylate. Once displaced this forms a glycosyl-enzyme intermediate that is unreactive and effectively traps the enzyme.

Chapter 9

9.1. Glycocode—the concept; glycolipids and glycoproteins; lectins; sophisticated signalling because of structural diversity, Examples—asialoglycoprotein receptor, legume nodulation and Nod factors, neural cell adhesion molecule and heparin sulfate; CD4-HIV and gp120, inflammatory response.

9.2. There are far too many different possible permutations to even consider making all possible variations (e.g. $> 1.05 \times 10^{12}$ for hexamer).

9.3. Flattened ring (e.g. by unsaturation), side chains that can bind to enzyme pocket, positive charge to mimic transition state positive charge development. Once you've done this look up the Glaxo-SmithKline anti-flu drug Relenza^{TM}, a potent sialidase inhibitor.

9.4. In fact NBDNJ is thought to inhibit the glycosyltransferase that puts a sugar on the ceramide lipid part. Ironically it seems as though the butyl chain mimics the hydrophobic chain and the DNJ part mimics the ceramide turn portion and neither mimics the sugar part!

Further reading

Monosaccharides: Their chemistry and their roles in natural products. P. Collins and R. Ferrier 1995 John Wiley & Sons, Chichester: This book is excellent and develops in detail many of the themes started here.

Preparative carbohydrate chemistry. S. Hanessian, (ed.) 1998 Marcel Dekker. This book's strength is that it contains some useful experimental details.

Molecular and cellular glycobiology. M. Fukuda and O. Hindsgaul (eds.) 2000 OUP. A nice account of the current state-of-the-art in glycobiology.

For reviews of chemical oligosaccharide synthesis see: H. Paulsen, *Angew. Chem., Int. Ed. Engl.*, 1982, **21**, 155; G. J. Boons, *Contemp. Org. Synth.*, 1996, **3**, 173; G. J. Boons, *Tetrahedron*, 1996, **52**, 1095; K. Toshima *et al.*, *Chem. Rev.*, 1993, **93**, 1503; B. G. Davis, *J. Chem. Soc., Perkin Trans. 1*, 2000, 2137.

For reviews on enzymatic oligosaccharide synthesis see: C. H. Wong *et al.*, *Angew. Chem., Int. Ed. Engl.*, 1995, **34**, 521; V. Kren and J. Thiem, *Chem. Soc. Rev.*, 1997, **26**, 463; D. H. G. Crout and G. Vic, *Curr. Opin. Chem. Biol.*, 1998, **2**, 98; Chapter 5 in *Enzymes in synthetic organic chemsitry.* C.-H. Wong and G. M. Whitesides 1994 Pergamon.

For further reading on glycoconjugates, glycocode, communication by carbohydrates and carbohydrate drugs see: J. C. McAuliffe, O. Hindsgaul, *Chem. Ind.* 1997, 170; B. G. Davis, *Chem. Ind.*, 2000, 134; B. G. Davis, *J. Chem. Soc., Perkin Trans. 1*, 1999, 3215.

For more information on glycobiology see: A. Varki, *Glycobiology* 1993, **3**, 97; R. A. Dwek, *Chem. Rev.* 1996, **96**, 683.

For special issues on carbohydrate science see:
Chem. Rev. 2002, **102**, Issue 2 (February);
Chem. Rev. 2000, **100**, Issue 12 (December);
Science 2001, **291**, 2337–2378.

Index